Hallucinate!

An Introduction to Non-Existing

Damian Timev

To my daughters and to their mother

Hallucinate!

An Introduction to Non-Existing

Damian Timev
To my daughters and to their mother

6TH BOOKS

London, UK
Washington, DC, USA

CollectiveInk

First published by Sixth Books, 2025
Sixth Books is an imprint of Collective Ink Ltd.,
Unit 11, Shepperton House, 89 Shepperton Road, London, N1 3DF
office@collectiveinkbooks.com
www.collectiveinkbooks.com
www.6th-books.com

For distributor details and how to order please visit the 'Ordering' section on our website.

Text copyright: Damian Timev 2024

ISBN: 978 1 80341 356 3
978 1 80341 357 0 (ebook)
Library of Congress Control Number: 2024921256

A CIP catalogue record for this book is available from the British Library.

Design: Lapiz Digital Services

UK: Printed and bound by CPI Group (UK) Ltd, Croydon, CR0 4YY
Printed in North America by CPI GPS partners

We operate a distinctive and ethical publishing philosophy in
all areas of our business, from our global network of authors to
production and worldwide distribution.

Contents

Contents

Acknowledgements

Many thanks to Dar for her comments and corrections. Thanks to Damana and Timana for the many chats on the subject of *paranormal*.

Thanks to Annie Wilder for guiding me through the world of publishing.

And thanks to Sarah J. Hodder for her excellent copyediting.

Introduction

In the year 1973, eight researchers led by David Rosenhan checked into eight different hospitals and declared that they had "heard voices," and that that was their single medical concern. Seven of them were immediately diagnosed as schizophrenics, while the eighth was diagnosed with "manic depressive psychosis." On the basis of their single declaration, even though they behaved normally, they were hospitalized for up to two months.[1]

Hearing voices that do not originate from our reality is a type of hallucination; the technical term is "auditory verbal hallucination." It is a fairly common phenomenon. Still, as Rosenhan's experiment strongly suggests, many psychologists consider it a manifestation of mental or physical pathology. We will most decisively not follow that line of thinking.

In fact, our narrative will head in altogether an opposite direction: we will argue that responsibly inducing hallucinations enhances the scope of our perception and sets us out to a mystic-like way of comprehending realities beyond corporeal perceptions.

We expand on our apologia further. First "hallucinating" is our mode of existence — deep down, we hallucinate all the time. Secondly, the reality that we perceive does not correspond to the objective reality. We hallucinate a universe into a subjective existence. This subjective universe is only partially common to all people; different individuals experience different "objective" or material realities. It is then only natural to explore the tentativeness of the so-called objective reality by trying to deliberately alter its perceptual output. Thirdly, and perhaps most importantly, by hallucinating we are exploring the relationship of the perceiver and the perceived, thus looking for ways to solve the puzzling enigma of "I". One of our main goals of this project is to clarify and elaborate on these ancient postulates.

We will start with the common definition of hallucination and we will use it to classify a set of tentative subclasses of hallucinations. Eventually we will be forced to make a profound change of meaning — at some point it will be obvious that hallucinations are not merely fantasies of delusional individuals. They are much, much more — they are consequences of tangible changes of parameters that define the relationship between the perceiver and the perceived reality. In other words, we hallucinate realities that are as objective (or non-objective) as the normal reality, which, in itself and in a way, is a large hallucination. I am, of course, aware of the gravity of these claims.

We will combine many anecdotes and controlled laboratory experiments, starting with simple examples and ending with outrageous cases of hallucinations. However, our aim is not to give an anthology of the phenomenon, but rather to provide sound support for our postulates. The hundreds of examples that we provide, and the millions that exist, yield a very high-resolution global picture of a reality far beyond the common materialistic paradigm.

We allow the possibility that we have misinterpreted some of the cases we consider. That does not change anything of substance. Changing the color of a few pixels will barely affect this picture of millions of pixels that we are outlining.

"Extraordinary claims require extraordinary evidence," they say. Extraordinary evidence that the ordinary materialistic paradigm fails miserably does exist. There are billions of counterexamples. For the gatekeepers, for the social manipulators, and for the unsuspecting followers of the materialistic religion disguised as science, that does not matter. No evidence is extraordinary enough for them; billions of counterexamples notwithstanding, they will label you as being unscientific and irrational for claiming otherwise. We will disregard the forces of deception, and we will keep going where the facts lead us.

Every book is a personal account. One can mask emotions and points of view in some cases by invoking rationality. But even in the most formal theories one must start with axioms, and the choice of these — which must be made at some point — is often a reflection of one's personality. We will not be very formal. I will not hide my own experiences and my own emotions.

We will have to abandon conventionality, and we do it with some pleasure. After all, "*Conventionality, my friend, is one of the worst forms of vanity because it is insidious.*"[2] It is also a great obstacle to wisdom, which I do not claim to possess. In our pursuit, at some point, we will have to face the inevitable: that almost everything we treat as normal is false, while arcane and mystic sources contain deep truths.

I never liked words. They are too discrete to faithfully convey the continuity that thoughts possess. But I do have something to say. So, I will say it.

This book is for the reluctant hallucinators among us. And that is the majority of humanity!

Chapter 0

Preludes

Prelude 1: Tanya

For a year, between the fall of 1981 and the fall of 1982, I was a conscript, serving as a peacetime soldier. That year was a unique experience for me — I acted mostly as a happy-go-lucky simpleton, without really being self-conscious about it. Fortunately, there were no obvious irreversible physical consequences of my idiotic tenure. I had a close call — only a few years later a vicious war raged over the same general area. At that time destiny had already taken me far away, to the other side of the ocean.

To be an idiot is blissfully addictive, and if the circumstances are favorable, one can carry on in that state for a lifetime. As a soldier for a year, one defers decisions to others. Higher-ranking military personnel give dumb orders, and you follow them with unobstructed devotion. It is a strict regimen that one must comply with in order to survive mentally. Asking sane questions, or doubting, risked failing the stupidity exams. The make-up exams were usually much more challenging; as they say, stupidity knows no limits.

Entering into this state of carefree dumbness was not instantaneous; as in cases of other altered states of consciousness, one needed some practice to achieve it. There were many exercises suitable for neophyte soldiers. I was particularly fond of the following drill: during the army drills on the mountains, the common test for affirming adherence to the doctrine of idiocy consisted of the junior lieutenant barking "Atomic to your left!", to which we — the common soldiers — were expected to increase the distance from the exploding mega-bomb by jumping on the ground to our right, and to lie

1

there for about a minute. As an added jollity most of the time the commander, who, having achieved immunity to nuclear bombs, stood indifferently during this drill, would choose a few mud puddles for us to dive into. At the end of the minute, he would invariably yell, "Danger over!" and we would stand up and continue with our merry, idiotic lives. So, I was happy.

My happiness at the time was compounded manifold by my wife. In her early twenties she was the most gorgeous woman in the world. I am saying this with total objectivity. After all, she was once crowned the most beautiful girl in a youth camp of some 500 teenagers, where, unbeknown to them, all girls were competitors. I loved her sky-blue eyes, her casual elegance — she used to be a ballerina in her childhood — her soft Slavic goodness curiously combined with the uncompromising frankness of the Northerners, her wild character, and her extreme intelligence. Her intelligence notwithstanding, what else could a common soldier have wished for? And, of course, I loved when she said that she loved me. For I knew she meant what she said.

Oh, how happy I was. For a day before she would visit me — a visiting wife was worth a weekend off — my face would freeze into a permanent smile, and nothing, nothing could have wiped it off. The whole garrison envied me. Idiots with beautiful and loving wives are the happiest people on earth.

On December 16, 1981, our daughter Tanya[1] was born. In the army's books that was not a good enough reason to be granted leave. So, I lingered in the barracks, immersed into thoughts about my daughter. What does she look like? Is she beautiful like her mother? How would it feel to hold my own child in my arms?

During the morning of December 18, I was handed an ominous telegram: there was some unspecified problem with Tanya's health and I was told to come as soon as possible. I managed to convince the corporal in charge to write a pass for me, ran to the train station, hopped into the first train going in the direction of my native city, and was on my way. There may have been other people in the dilapidated third-class carriage with un-upholstered wooden seats, but I did not notice them. I was only aware of the darkness that started enveloping my mind. I tried to brush it off but it would not go away. Something was terribly wrong.

When I arrived, I was met by my uncle and my brother-in-law. I was told that Tanya had passed away during the night. I cried.

I feel like crying now, as I write this, 35 years later, for my sadness for Tanya has not diminished with the years. On the contrary. At that time the events swept me like a torrent; I was too deep in the streams of life to try to make sense of them. It took me years to understand. Now I know what I was deprived of. I did not see her eyes. I did not hold her in my arms. I did not watch her smile. And, most importantly, I did not tell her that I loved her.

It will forever be my and my wife's personal tragedy. However, now I can see it from another angle. For something strange happened during the night between 17 and 18 December 1981, that eventually, some 20 years later, forced me to rethink my original, normal interpretation, not only of Tanya's short life, but of my own life too.

Prelude 2: Tanya, Not in My Dreams

One can only learn army life during peacetime by experiencing it. No book can convey the strange metamorphosis from a normal person into one who is conditioned to kill and be killed. It is a subtle change, not easily noticeable while it's happening; it consists of getting accustomed, in minute degrees, to the lightness and ephemeralness of life. This lightness of living is not being experienced as a philosophical concept, but rather as a state of existence. It is a descent — or an ascent — into becoming utterly insignificant, a fly. And once you are a fly, so is everyone else.

To make a virtual fly out of a normal person is not a slight accomplishment. Beside stifling the rational mind and — as we have noted — inducing stupidity, in order to achieve insignificance, the emotional components need to be reduced to very few, often dominated by absurd happiness that dumbness brings about.

The physical body also needs treatment to metamorphose into a good, expendable soldier. This was achieved through exposing it to misery. Miserable food, miserable hygiene — showers only every couple of weeks or more — and miserable healthcare, especially dental care.

Miserable breathing air too! No words can express the shock of the first inhaled breath, upon diving into the unventilated army dormitory, saturated by a putrid bland of dissipated sweat, intestinal gasses, and stink emanated by rotten teeth and feet fermented daily in waterproof army boots. It was a brand new atmosphere; it might as well have been from another planet. It usually took some 15 minutes to equate the air in the lungs with the air in the room, and to fool your body into believing it is normal air that runs it.

I don't know if it was because of some anesthetic effects of that stench, the bromine-laced food that we were fed, or simply because of the physicalness of daily life, but I really slept like a log that year. And, as far as I could learn from my comrades, this sound sleep was a rule much more than an exception. I would jump on the top level of my spring bunk bed, cover myself with the blanket, fall asleep in seconds, and the next entrance into waking life was when I would be roused early in the morning by the corporal in charge shouting, "get up army."

Only once in some 300 nights was my sleep interrupted without an obvious physical cause. And that happened the night Tanya died.

That evening was ordinary — I hopped onto the bed and fell asleep as soon as I pulled the blanket over my body. The sleep was normal until sometime in the middle of the night when I heard a long, anguished, terrible scream that lasted some five seconds. It was not at all dreamlike — I perceived it as a physical sound. It pierced my sleeping mind, and it made me sit up with a start.

I was not scared as I sat there in the dim light; I was merely confused. What on earth had just happened? What did I hear? Where did it come from?

For a few seconds I sat bewildered. My initial instinctive reaction was to brush the experience away, but it was too real for a quick-fix solution. And then a morbid thought entered my mind: something terrible had happened to my baby.

I sat motionless for a few minutes. I could not understand my emotions — there was a feeling of utter hopelessness that was new to me. It seemed as if something invisible had just snatched a chunk of my soul, and that there was nothing I could do to retrieve it.

The feeling lasted for a couple of minutes, after which the rational mind began its normalizing routine — it must have been a nightmarish dream, or a hallucination of some sort. It is plain silly to let superstitious thoughts harass me. Forget it and sleep well.

And so, I lied down again and let sleep take over.

The next day, when I got the ominous telegram, I knew within my heart what it meant. But my indoctrinated brain of the time handily won the battle. It dismissed the night event as a coincidental hallucination and refused to relate it in any way to "real" life. Thinking about this now I see that there was no alternative. I had not a shred of an idea about anything beyond the material world. I was raised to be, or brainwashed into being, a hard-core materialist. Any hint of spirituality — religious, mystic, or even rational — was, as far as I was concerned, a ridiculous mumbo-jumbo indulged in by the superstitious, or by weaklings who needed emotional reassurance in order to cope with the inevitability of death and annihilation.

My views have changed.[2]

Prelude 3: Another Hallucination and a Dream

There are two classical ways of interpreting Tanya's cry: it was a dream, or it was an auditory hallucination. As opposed to dreams, hallucinations happen in waking states, or at least in the intermediate states between sleeping and being awake. We will consider these notions more thoroughly later on. The fact that there was that strange coincidence of the event happening at the same time as the passing of Tanya must then be dismissed as being exactly that: a coincidence.

I will eventually argue that there was much more than a mere synchronicity going on that night. It will take a few chapters to make a reasonably convincing, yet still an incomplete argument.

For the time being we will consider two more events in my life. First, Tanya's cry was not my first auditory hallucination. And second, the "coincidence" happened to me again, many years later, in strangely similar circumstances.

When I was a teenager, I lived with my family in a first–floor apartment of a regular working-class apartment building. These were the times when there was no daily TV broadcast, and when personal computers and other techno-gadgets for solitary entertainment were a few decades in the future. The days with nothing to do were plentiful, especially when it rained outside. I don't really know why, but during one such "nothing-to-do" day I decided to play a game with my mind. I closed my eyes, and let my consciousness go inward. Of course, I did not know what "going inward" really meant, and so I did not know what I was doing. I simply emptied my mind and let it take its own course, not thinking about anything in particular. It was surprisingly easy — as if my real self found a well-trodden path from ancient times. I felt as if I was descending into a dark boundless cave.

I was at ease, and I just let the descent continue. And then, out of nowhere, I heard a cacophony of terrible cries. It was like a wall into which I was thrust at full speed. It scared the hell out of me, and I promptly scrambled back into the familiar world of normal perceptions. It was then and there that I decided never to repeat the experiment.

Now I have ambivalent feelings about that decision.

On one hand I feel sorry that I chickened out so easily. I did not know at the time that I might have travelled briefly along the so-called mystic paths, and I was, of course, completely ignorant of the fact that my encounter with the "demonic gatekeepers" was a fairly typical neophyte mind-explorers' experience. Mystics, shamans, experimenters with drugs (including professional researchers) know about these spiritual barriers that seemingly exist to keep the weak-hearted or unprepared from accessing the deeper, super-perceptual realms. At times, so do people who meditate regularly. For example, the medical oncologist and writer Slawek Wojtowicz, while meditating, encountered — in his mind (where else?) — a dark, "hostile" figure that paralyzed him for a few moments. He promptly abandoned that type of meditation. Years later he explained his experience as a typical encounter with a Threshold Guardian, and, like me, regretted his decision to terminate the meditations-induced inner-states journeys.[3]

Perhaps I might have had enough talent to reach the mystic realms by sheer deliberate intent. I will never know — and that I regret. On the other hand, my impromptu decision — to the extent it was *my* decision — never to dive again into the depths of my self was certainly a wise step at the time. Had I persevered and probed my inner consciousness further, I might have ended up in a lunatic asylum, where the standard "cure" for suspected schizophrenia at that time and in that society consisted of erasing the sufferer's mind and personality by electroshocks. The danger for me was not in going crazy, but rather in actually finding something interesting, or something extraordinary within the world beyond basic perceptions, and then failing to keep my mouth shut.

The other experience that I had and that was related to Tanya's cry happened many years later. In itself, it was just a dream, and so it could not be classified as a hallucination. However, it was an unforgettable and a remarkable dream. My sleeping mind was taken to, what appeared to me to be, a kind of a heavenly realm, where I was immersed into an ocean of the most beautiful, amazingly rich, living light. Waves upon waves of nourishing and loving colors bathed my mind! I experienced a complete bliss, and a feeling of unbounded happiness. There was also a lingering, puzzling sense of comprehending something fundamental.

I was so happy that I was still smiling broadly when my teenage daughter woke me up in the middle of the night. She was crying. Her grandmother, my mother, had died that night, and it was my daughter who picked up the phone and was told of the sad news.

So, I sat on the edge of my bed trying to understand for a few moments if I was happy or sad. The most antipodal, contradictory emotions seemed to have coexisted within me for a brief period of time. In a short while, the (so-called) real life prevailed and the realization that my mother was gone steered my emotions in the direction of sadness.

My mother passed away in Macedonia, my native land, an ocean and large chunks of two continents away from me. She had a heart attack, and so the timing of her death was unexpected. Considering this, here we have the coincidence of two ostensibly unrelated, yet apparently linked events happening at the same time: my mother departing for the so-called heavenly realms, and me being taken there in my dreams.[4] I now firmly believe that it was not a coincidence at all. The death of the very person whose love for me, her first child, was the purest that earthly love could be, and the dream of "heaven" happening at the same time is not a fluke; a far more reasonable explanation is that it was a message.

I mention in passing that accounts of loving, beautiful light experienced at the moment of death are ubiquitous among people who have had near-death experiences. Have I witnessed the entrance of my mother into the postmortem realms? And, in Tanya's case, have I heard the anguished cry of my first child? I am not looking for definite answers at the moment, for I know they are not easy to justify. I only want to say that the questions are reasonable.

My assessment — that my dream and my mother's passing are likely related events — is not just emotional. It is also based on evidence, both anecdotal and obtained by means of scientifically controlled experiments. We will cover both. Regarding the former, there are hundreds, perhaps thousands of *recorded* dreams or visions ("coincidental hallucinations") of geographically distant, but emotionally close, dying people. There are so many veridical cases of such sort that it is statistically hugely unlikely that the explanations invoking coincidences are even remotely viable.

We will come back to this theme later and we will consider some of the evidence. The cases and the studies we present here lead us to extravagantly simple models of the deeper reality. This extravagance does not bother me at all — reality is extravagant. It is also the simplest possible — an unchanging constant. We will elaborate; it will take some time.

Chapter 1

Hallucinations: Definition, Types, and Scope

千里之行, 始於足下

A journey of a thousand miles *begins with a single step.*

1.1 On Defining Hallucinations

What are hallucinations? The prominent neurologist, Oliver Sacks, defined them as follows: "... hallucinations are defined as percepts arising in the absence of any external reality — seeing things or hearing things that are not there."[1]

This definition is fairly standard in the sense that other classical definitions are essentially minor modifications. For example, instead of "external reality," other psychologists refer to things that *really exist*, as opposed to ostensibly intangible mind products. *Diagnostic and Statistical Manual of Mental Deceases*, published by the American Psychiatric Association, defines hallucinations as "sensory perceptions without external stimulation of the relevant origin". It is understood that the origin of sensory perception is *relevant* if it is sourced in the material world.

The explicit or implicit reference to the "(material) reality" in the common definitions of the notion of hallucination is remarkable! It is the pinnacle of human audacity to define something — anything — in terms of reality! Besides the obvious circular reasoning — for the so-called external reality for one person includes any other person's individual experience, hallucinations and all — the main problem is in the grandiosity of the notion. How on earth can one define anything in terms of such a grand notion as *reality*? For, what is reality? Or, assuming "external reality" refers to the material world, what is matter? Do we know the answers? Has our current science established the objectivity of matter? And, moreover, is the texture of matter substantially different from the texture of hallucinations?

We will pay attention to these questions in a separate chapter. For the time being we note that the hypothesized objectiveness of matter in the psychologists' definition of hallucinations is not a universally accepted scientific postulate. The problem is with the dual behavior of the elementary particles, comprising the units of matter. Numerous scientific experiments confirm that at a micro level matter behaves as energy (or quantum waves), and as "solid" particles, and that the latter happens only when these elementary particles are being *observed*.

The question of bare existence of the basic particles of matter is meaningful. Modern physics, especially quantum mechanics, does not always support, and in some cases even contradicts the materialistic paradigm of objectivity of material world. The quantum physicist, Casey Blood, flatly stated: "After years of poring over the evidence, however, I have concluded that there is more reason to believe that particles don't exist than that they do."[2]

If elementary particles "don't exist," then neither do we. Is this a far-fetched postulate? Why should it be: if a particle does not exist, then what is wrong with asserting that a conglomerate of a few trillions of such particles also does not exist? Is our reality a quantum reality?

Turning a page and considering the same question from the mystic initiates' point of view, the answer is virtually unanimous: the material reality is so plastic and malleable by thoughts that it could be said that it is illusory. There is a curious anecdote narrated by the psychologist, Jack Kornfield, regarding his Tibetan spiritual teacher, Kalu Rinpoche. Kornfield wanted the Tibetan to describe for him in one sentence the essence of Buddhism. The Tibetan replied: "I could do it, but you would not believe me and it would take many years to understand what I mean." The psychologist insisted, and the Tibetan's response was: "You do not really exist."[3] Kornfield abandoned the practice of traditional psychology and became a Buddhist monk. It took me years to convince myself that I understand Rinpoche's momentous statement. We will come back to it by the end of our story.

Do objects that are hallucinated exist? If not, is their non-existence substantially different from the non-existence of micro particles of matter, and from matter in general?

One of our theses is that the perceptional reality of what we call matter is not that different from hallucinated percepts. We will elaborate on this later.

For a while we will tentatively and grudgingly accept the standard definition of hallucinations, and we will proceed from that point. We will assume that there is an objective material reality, external to our consciousness and independent from it.

For practical reasons, we will also use the following working definition of hallucinations: if you experienced a phenomenon that defies the basic tenets of material "reality", and if you report it as such to your run-of-the-mill physician, psychologist or psychiatrist, then the event is a hallucination if you are diagnosed as manifesting symptoms of anything between proto–schizophrenia and being a total nutcase. Every event that we will describe in our narrative falls within the scope of this definition. We will start with relatively innocuous cases of dreamlike hallucinations ("symptoms of schizophrenia"), and we will steadily go in the direction of more extravagant cases — say, (Section 6.5) you entered a deserted mine shaft, and upon exiting, the scene changed, and you are greeted by two levitating humans, bewilderingly staring at you ("symptoms of having lost all of your marbles").

We are, of course, not claiming that all medical professionals fit one mold. There are many among them who are aware of the fundamental difference between a mystic event and a psychotic episode.

Eventually we will be forced to revisit the problem of properly defining hallucinations. Actually, at one point we will need to face the modified problem — what is **not** a hallucination? The answer that will be suggested by our narrative will be very few things!

1.2 Types of Hallucinations

There are numerous classifications of hallucinations, depending on the criterion that is applied.

The most common subdivision of hallucinations uses our basic senses of perception as the main criterion: there are visual hallucinations, auditory hallucinations, including auditory verbal hallucinations (perceived as sounds), tactile hallucinations (perceived through touching), olfactory hallucinations (perceived as odors), and gustatory hallucinations (perceived as taste). The first two seem to be the most common.

There are hallucinations that combine some of the above five types. The most interesting and the most intense are the ones combining the first three: seeing, hearing and touching.

Hallucinations — keep in mind that we are, for the moment, using the psychologists' definition based on the intuitive notion of "reality" — can be induced by meditation or by any kind of suppressing or dulling of the carnal perceptional functions. The simplest — but not the easiest — such method is by repeating ad nauseam a single mundane action: for example, by clapping two stones, as in some shamanic initiations, or by repetitive chanting or praying, as in some religious rites.

The most common spontaneously experienced hallucinations of healthy people happen on the boundary between being asleep and being awake. There are hypnagogic hallucinations happening at the moments of transition from being awake to being asleep, and hypnopompic hallucinations at the moment of waking up. According to some studies, a large proportion, if not the majority of adults, experience at least once during their lives a hypnagogic or a hypnopompic hallucination.[4] For example, McKellar (1957) found that of 182 students [in his experiment] 115, or 63.16%, reported hypnagogic imagery, whilst 39, or 21.42%, reported hypnopompic imagery.[5]

There are hallucinations caused by illnesses, injuries, or by anything that damages the normal functioning of our bodies. The common medical interpretation of the phenomenon is fairly straightforward: sickness causes stress, our mind is blurred, perceptions are impeded, and we are prone to delusions of various types, including hallucinations. It is a rather simplistic attempt to explain away the problem, and it does not account for numerous veridical anecdotes of sick or dying people hallucinating events that had really (!) happened and of which they had no *normal* knowledge.

A non-pathological — and non-medical — explanation starts with our premise that hallucinations are not that different from percepts of the material world. However, as in the case of deliberately induced hallucinations, some kind of suppression of the basic senses is necessary in order to hallucinate, which is precisely what happens when one is sick. From this point of view, we can say that sickness sometimes enables us to access other realities. In this context, especially prominent are the so-called deathbed hallucinations which deserve special attention (see Chapter 3).

Consuming hallucinogenic drugs, also known as psychedelic drugs, causes, of course, hallucinations. After all, this escaping to other perceived realities is usually precisely the reason why many people take drugs. Both experimental and anecdotal evidence imply that the simplistic explanation that hallucinations in such cases are merely mental plays of deluded minds is false. As we will see there is more to it — so much more that using the neologism *psychonauts* for some experiencers is rather apt. In many cases psychonauts sometimes access other consensual or shared realities, and consequently their experiences cannot be explained away as random individual fantasies!

Another large group of hallucinations consists of those caused by hypnosis or self-hypnosis. We will cover these too; there are very interesting cases among them.

One such case is described in Da Silva Mello's book *Mysteries and Realities of This World and the Next.*[6] In an experiment performed in the early 1930s, Ludwig Mayer hypnotized a boy and told him that he was going to be able to talk to his father, who had been dead for many years. A few moments later, the boy opened his eyes, which were full of tears, and asked, "Is that you father?" Then they talked for a short while, when Mayer, "in order to avoid too much suffering for the boy," brought him out of the hypnotic state. Mayer, of course, saw nothing of the father, and took it for granted that the boy was imagining the whole episode. The boy on the other hand, was adamant in his claim that he did talk to his father. He was so much convinced of the reality of the encounter with his father, that a subsequent counter-hypnotic suggestion by Mayer to bring the boy out from his illusion was not successful — the boy still insisted that he talked to his father. Mayer took it for granted that the boy properly hallucinated and was imagining things that were simply not there. In this book we will take the boy's side and head in the direction of an alternative explanation: the boy's spectrum of perception was altered through hypnosis, and he did see the "ghost" of his father.

This will bring us closer to one of our main queries: is the *normal* reality nothing more than a consensual perception through a kind of huge mass hypnosis or self-hypnosis? If yes, then what exactly is the source of the illusion, and what are the parameters of the underlying phenomenon? Who hypnotizes us? Who exactly self-hypnotizes us? Furthermore, who is "us"; what is "self"?

Very interesting and consequential hallucinations happen during near-death experiences (NDE). Related to these kinds of hallucinations are out-of-body experiences (OBE). Again, there are exquisitely interesting cases of such hallucinations, and many do not allow orthodox explanations. We will devote a chapter to these (see Chapter 4).

Group hallucinations are especially relevant for what we are trying to model, since they are not easy to explain away. It is possible to stretch the scope of conventional psychology and look for orthodox explanations in some trite cases of group hallucinations. The recipe usually combines delusion and suggestibility and looks reasonably at the weak end of cases of group hallucinations, for example cases of small numbers of people seeing a ghost.[7] But what about when hundreds of people simultaneously see something that is provably not based on normal reality? And what about interactive group hallucinations, when groups of "real" people have a chat with a group of "unreal" people? And what about groups of people wandering into a world out of the *normal* time and space paradigm? Should we dismiss all such cases as phony? There is not much of a choice except doing precisely that, if we insist on staying confined within the frames of the orthodox models of "reality". That is what we will most emphatically not do.

There are waking hallucinations. You are in the kitchen frying eggs for breakfast, and you hear a human voice, but there is no one around. Or, to give a more extreme example, you encounter a full-bloodied replica of yourself, and perhaps use the occasion to exchange a pleasant chat. These types of hallucinations have been reported many times. Psychologists, in their constant quest to classify what they deem to be pathological states of human mind, use the name *autoscopy* for episodes of encountering oneself. Most of the time autoscopy is associated with out-of-body experiences (OBEs). A larger class of hallucinations is covered by the term doppelgänger — in such cases a double of a living human manifests to other people. If, at the same time, someone sees the "original", then this becomes a case of bilocation. There is considerable overlapping between these two terms, and we will not distinguish them. There is an abundance of reported cases of these types. Later I will add my own simple case of a doppelgänger to this collection. Waking cases of hallucinations, when experienced by healthy individuals, are harder to dismiss off hand, precisely because the dulling of the physical senses by sleep or otherwise is not present.

On the other end of the spectrum are the feeblest hallucinations, so feeble that they are called pseudo-hallucinations. Pseudo-hallucinations "lack the substantiality of perceived external stimuli."[8] In other words, you perceive something that is within your "inner subjective space," and you are aware of its illusory nature during the episode. Some psychologists expand this notion to also cover hallucinations that are perceived as external, but are *a posteriori*, after some deliberation, declared as actually residing in our brain or mind, and only there. In fact, according to many psychologists every hallucination is a pseudo-hallucination in that sense, and if you think otherwise, you are delusional, hysteric, schizophrenic, or mentally screwed in this or that way. The litmus test for being sane is to accept the verdict of the scientific establishment and damn you if you don't.

In the next chapter we will start our main narrative with minor hallucinations — say, you see-dream-hallucinate a flower that does not actually (*sic!*) exist. Here is an example of a minor visual and auditory hallucination, ending with an unusual veridical twist:

I was awakened one morning with a vision of a woman's forearm holding a box, translucent. And, in the box, there was a beautiful white gardenia. And it wasn't the type of gardenia that we see in this world; it was a spiritual flower. And I heard a voice just as clearly as my own saying, "Take this flower, take this to Mrs. Henry, my mother, and tell her I am always with her." Now, Dr. Ring, I didn't know any Mrs. Henry, but I had the habit of going to the corner of the cafeteria [at work] every morning for a cup of coffee and I sat at the counter. And I was the only person there except for a woman that sat at the opposite end of the counter. There was no one else there but ourselves. And I heard the waiter say to her, "Would you like another cup of coffee, Mrs. Henry?" And I said, "Do I dare?" A perfect stranger! I went up to her afterward and I said, "I beg your pardon. Your name is Mrs. Henry?" "Yes." "May I tell you something?" "Yes." And I told her what I had heard. She looked at me with stricken eyes, and she said: "A gardenia was my daughter's favorite flower and she has just been killed in an automobile accident."[9]

After covering minor hallucinations, we will progress in the direction of far more interesting cases — for instance, when you wake up with a dreamed flower in your hand. Here is the poet Samuel Taylor Coleridge:

What if you slept?
And what if,
In your sleep
You dreamed?
And what if,
In your dream,
You went to Heaven
And there plucked
A strange and

Beautiful flower?
And what if,
When you awoke,
You had the flower
In your hand?
Ah, what then?

Forget flowers! The story about hallucinations is far more extraordinary. How about you find the pharaoh, Seti the First, in your bed when you awoke? This allegedly has actually happened (Section 8.4). Allegedly? What is it that does not happen allegedly?

Coleridge, by the way, was no stranger to hallucinations; in his case they were likely induced by taking opium. Here is a quote from *Romanticism, Medicine and the Natural Supernatural* by Gavin Budge: "Coleridge interprets hallucinations as a 'symbol' of a greater truth about the immaterial nature of perception, rather than dismissing it as a mental aberration."[10] The quoted statement comes close to my own view.

If we accept our final verdict about what hallucinations really are, then this should be a book about everything. We are forced to stay focused and for a while apply the standard definition as a criterion for hallucinating. This is sufficient and will lead us to where we want to go anyway.

We must mention that the term "hallucination" has been consistently misused as a kind of denigrating label for any experience that does not fit the sanctioned and controlled narrative. Take, for example, close encounters with UFOs and (extraterrestrial, non-human) alien intelligences. There is not a shred of a doubt that the alien phenomenon, extraterrestrial or terrestrial, is, in general, authentic. Not a shred! Yet the experiencers are often labeled as hallucinators or weirdos. Thus, this important issue is cynically relegated to obscurity by the narrative-controllers and their naïve front-stage scientists. Since encounters with non-human intelligent entities are not hallucinations according to our tentative definition of this term, we will not cover them. However, there are cases when phenomena of large-scale hallucinations and non-human intelligences are intertwined; these will be mentioned.

We stay with this subject for a little while longer. Orthodoxy and the social-mind controllers would claim that the universe is sparsely inhabited by intelligent human-compatible beings, and that, as a consequence, we are alone in our corner of the galaxy. They search for signs of intelligent extraterrestrial life by taking distant peeks through sophisticated telescopes or other instruments. Distant? Should we not first settle the problem of the nature of matter? For there is no conventional distance in a *real* world without *real* matter.

1.3 Hallucinators: A Brief Historic Overview

Hallucinators — a necessary neologism — are people who hallucinate. Who are they?

Hallucinators are us! It seems safe to claim that the majority of humanity, past and present,[11] has experienced some kind of hallucinations. Today, under pressure from those who tell us what is real and what is not, most of us tend to dismiss hallucinations as inconsequential, and in many cases we degrade them to dreams or waking fantasies.

It used to be very different — hallucinations were visions, glimpses into other realities. A part of it was certainly due to superstition. But that is not the most important aspect of the phenomenon. We will argue that it is reasonable to postulate that the perceived reality in pre-materialistic times was *objectively* much wider. Not very long ago ghosts and fairies were universally acknowledged and socially accepted ingredients of our common reality.

Consider, for example, the Celtic people of the British islands. Up to the nineteenth century their world teemed with various types of fairies. Evans-Wentz in his *The Fairy Faith in*

Celtic Countries, writes: "So firmly did the old people believe in fairies then [early nineteenth century and earlier] that they would ridicule a person for **not** believing."[12]

What caused the demise of the perceived Fairyland? Here is what an old man from the Isle of Man told Evans-Wentz: "Before education came into the island more people could see fairies; now very few people can see them." [13]

In Africa socially acceptable clairvoyant seers lasted some couple of centuries longer. When Carl Jung visited Uganda in the 1920s, he met a local shaman, *Laibon* in the local language, and asked him about his dreams. Laibon's reply was amazingly similar to the old Manx man's charge:

> *He answered with tears in his eyes, "In old days the laibons had dreams, and knew whether there is war or sickness or whether rain comes and where the herds should be driven." [...] But since the whites came to Africa, he said, no one had any dreams any more. Dreams were no longer needed because now the English knew everything.*[14]

Education eradicated superstition; materialistic education eradicated whole worlds and replaced them with a much narrower and more controllable realities. At a point in human history we were told that Fairyland did not exist, and the spectrum of our perceptions obligingly shrunk to match the directive. However, Fairyland doesn't exist in the same sense that we don't, as per the wise lama, Kalu Rinpoche.

We are not implying that the fairy-populated realities are superior or *better* than today's *normal* reality. Indeed, in some cases it is tough to prioritize — do we prefer regular contact with demonish incubi to the Inquisitorial burning of thousands of "witches", including children? We merely postulate that these alternative realities are as much real — or unreal — as the common, approved, certified world.

Certain types of hallucinations have always been ubiquitous. Take for example visual hallucination of dead humans, commonly known as ghosts. These are fairly common today and have been common in the past. Henry More stated in his book *The Immortality of the Soul*, published in 1642,[15] that, "The examples [...] of the appearing of the Ghosts of men after death are so numerous and frequent [...] that it may seem superfluous to particularize in any."[16] More's lament is to be repeated numerous times by many frustrated researchers into the so-called paranormal phenomena.

Going further into the past takes us to outrageous cases. The following also comes from More's old book — Agrippa [63 BC to 12 BC] was a Roman general and statesman:

> *Agrippa writes in Cretian [Cretan] Annals how there [...] the spirits of the deceased Husbands, would be very troublesome to their Wives, and endeavour to lie with them, while they could have any resource to their dead Bodies. Which mischief therefore was prevented by a Law, that if any Woman was thus infested, the Body of her Husband should be burnt, and his Heart struck through with a stake.[17]*

Now, this should be dismissed as superstition, a relic of the bygone ages. Or should it? The following comes from the book *Hallucinations*, by Oliver Sacks:

> *A general practitioner in Wales, W. D. Rees, interviewed nearly three hundred recently bereft people and found that almost half of them had had illusions or full-fledged hallucinations of a dead spouse.[18]*

We are not told what *full-fledged* means in this context. As we will see, the deeper reality is outrageously extravagant.

Rees also found out that only 28% of the almost 150 hallucinators that he had interviewed told anyone about their experiences![19] There is more than just social stigma and exposure to ridicule that forces people to keep these experiences to themselves. There is also a long and collectively remembered history of persecution.

People who had spiritual experiences of any sort, hallucinations included, have always been — and still are — a danger to the dogmatic religious establishments, and, at times, to ostensibly secular elites who found them more difficult to rule. It is not an overstatement to say that they have repeatedly been exterminated throughout the ages. Seers, spiritual initiates, clairvoyants, and people with *second sight*, all of them hallucinators galore, were especially targeted. It suffices to mention that many tens of thousands of women were executed in late medieval Europe during the 300-year period of witch-hunts.[20]

In modern times, a significant segment of the established scientific caste, manipulated by a thin layer of the Illuminati, scorned, denigrated and, indeed, persecuted those people whose perceptual spectrum exceeded the "scientifically" sanctioned one. Hallucinators were in this group. It is interesting that in the modern western world up to the beginning of the twentieth century, hallucinations were a legal criterion for judging someone insane.[21]

At times psychiatrists played prominent negative roles. Consider, for example, the legacy of the psychiatric researcher Franz Kallmann, who looked at genetic bases of psychiatric disorders, particularly of schizophrenia. In 1935, during the ascent and consolidation of the Nazis in Germany, he gave a speech in which he advocated forceful sterilization of *relatives* of schizophrenia patients. In 1939 he published a book, and on the last page he wrote that, "we may arrive in the not too distant future at a **complete solution** of the problems dealing with schizophrenia trait."[22] And "a complete solution" indeed arrived a couple of years later. Kallmann was one of the prominent scientists of the time who set up the pseudo-scientific stage for the extermination of schizophrenic patients in Nazi Germany — estimated numbers range between 100,000 and 137,500 — and sterilization of many more, including a two-year old girl. Ironically Kallmann had to flee Germany; he was one-half Jewish.

The reader should keep in mind Rosenhan's experiment (first paragraph in the Introduction) which showed how easy it was during the 1970s to be misdiagnosed as schizophrenic. You have hallucinated once, you are a schizophrenic, case closed. This was surely even easier in Nazi Germany.

1.4 Hallucinators: Commonness

Because of all of the above, it is fairly safe to postulate that scientific polls regarding hallucinations usually give the minimal numbers of hallucinators in the polled samples — there are many people who are very reluctant to admit or to share their experiences of hallucinations.

That notwithstanding, the percentages of declared hallucinators with respect to the total population are high anyway. Let's consider this further.

The first large-scale census on the subject of hallucinations was administered by the (London) Society of Psychical Research. In 1889 they asked 17,000 persons the following question: "Have you ever, when awake, had the impression of seeing or hearing or of being touched by anything which, so far as you could discover, was not due to any external cause?" [23] The affirmative answer was given by 1694 persons, or almost 10%.

We mention in passing that a follow-up survey of the 1694 hallucinators uncovered many veridical hallucinations, when the hallucinating experiences tallied with distant events that occurred at about the same time. We will come back to this issue in the following chapters.

Note that the survey of 1889 excluded hypnagogic and hypnopompic hallucinations. Turning our attention to these, we mention a study by McKellar in 1957: out of 182 students that he surveyed, 115, or 63.16% (!), reported hypnagogic hallucinations, and 39, or 21.42%, reported hypnopompic hallucinations.[24] These are large percentages, but the samples are moderate in size.

Also note that the last two studies are disjointed — hallucinations that were the subject of the 1889 poll were excluded in the 1954 study, and vice versa. However, since there are certainly hallucinators who have experienced hallucinations covered in both studies, we cannot merely add the respective percentages to get the percentage of all studied hallucinators relative to the whole population.

So far, even after we factor in statistical errors, we can only conclude that the proportion of hallucinators in the whole population is "large"; likely larger than 50%. Our modest goal in the remainder of this section is to bump up "large" to "very large." Let's keep going!

Let's consider polls regarding cases of autoscopy — seeing oneself from outside one's body. This class of hallucinations may have some overlap with the ones studied in 1889, but has probably near-empty intersection with hypnagogic/hypnopompic hallucinations.

In 1954, Hornell Hart asked 113 students the following question: "Have you ever actually seen your physical body from a viewpoint completely outside that body, like standing beside the bed and looking at yourself lying in the bed, or like floating in the air near your body?" 14 out of 42 students, or 33% of them replied "yes."[25] A couple of decades later Palmer and Dennis polled 534 townspeople and 268 students in Charlottesville,

Virginia, and asked them essentially the same question, but with an extra instruction: "If in doubt, please answer 'no'." The results were consistent with Hart's study: 25% of students and 14% of townspeople reported having OBE (and autoscopy), but this time without any doubt.[26]

Let's define "ghosts" as visually perceived manifestations of deceased people. Most of the "ghost" anecdotes are rather clichéd; there is not much wisdom that one can squeeze out of them. There are, however, very interesting cases among them; we will note them later. Moreover, the sheer ubiquity of the ghost encounters is meaningful. Here are a few polls and studies regarding such hallucinations:

- The work *Phantasms of the Living* by Myers *et al.*, contains some 700 cases of apparitions, obtained by canvasing 5705 persons, chosen at random.
- In an early study, Rees (1972) found that 13% of bereaved people, most of them aged over 60, had heard the voice of their deceased spouse. The longer the marriage had been, the more probable hallucinations of all types were. Only 28% had told anyone about their experiences.[27]
- In 1984 or soon after, the National Research Council conducted a study of experiences of contact with the dead by the general public. The results of this poll show that, out of 1473 people with deceased spouses, 67% reported such contacts, while a total of 42% of the general public makes similar claims.[28]
- According to a poll taken in 1987, 42% of American adults believe that they have been in contact with someone who had died, and 67% of all widows believe they have had a similar experience.[29]
- Interesting census results regarding contact experiences of people grieving over lost relatives

were collected by Dianne Archangel. The participants were first classified as "intuitive-feeling" (those who rely more on intuition and feeling in their life) and "sensing-thinking" (those who use their perception of the material world and judge it accordingly). 96% of the first category reported some after-death contact, and 100% of the second group reported nothing! (See pages 89–90.) This is related to the sheep-goat effect; we will come back to it later.[30]

- In a poll (Roper poll, 1991) conducted with 5942 people, "11% reported having seen a ghost."[31]

Finally, there is one large class of people not covered in any of the above polls: dying people. Ultimately that is 100% of us! There is good reason for this lack of coverage: dying people do not fill out census questionnaires; they are too busy dying. However, we do have some interesting results anyway.

During the 1950s, Karlis Osis conducted a pilot survey. He sent an 11-question questionnaire to 5000 physicians and 5000 nurses in India and USA pertaining to their observations of deathbed experiences by their patients.[32] He was particularly interested in the emotions and experiences of the patients who were fully conscious just before they died.

It turned out that about 2200 of the dying people were fully conscious during the last hour before passing over. A significant proportion of these people — 884 cases or around 40% — reported hallucinations of various personalities; the majority of these (52%) were images of dead persons. Interestingly, in four cases the apparitions were perceived collectively by the patient and by other people.

Especially relevant for our purpose are veridical deathbed hallucinations; we will cover these in more detail in Chapter 3.

1.5 Reality, Step 1: Hallucinations and Ubiquity

$$\frac{1}{\pi} = \frac{2\sqrt{2}}{9801} \sum_{k=0}^{\infty} \frac{(4k)!(1103 + 26390k)}{(k!)^4 396^{4k}}.$$

This intimidating formula, representing a relationship between certain numbers, does not concern us per se. What is interesting to us is the mode of its discovery: it was bestowed to the Indian mathematician, Ramanujan, by the goddess, Namaguri. She wrote it down for him during one of his many mathematical dreams. Or were they dreams at all?

Ramanujan did not interpret his visions of Namaguri as fantasies generated by his sleeping freewheeling brain — which is what the orthodoxy tells us dreams are. He was a full-blooded mystic who believed in Namaguri's *objective* existence. After all, he once spent three consecutive days and nights chanting Namaguri's name in an Indian shrine, ultimately managing to invoke her presence, and receive from her important advice regarding choosing a direction in his life. In modern terminology, that would be a case of an induced hallucination.

Dreams are often used by hallucinators as a kind of a safety net — you hallucinate your dead grandmother, you are excited about it and feel an urge to share your experience with your skeptical friend. You simply call it a dream, and no eyebrows will be raised. For, unlike hallucinations, there is no pathology associated with dreams. Dreams are considered normal and explainable.

Our materialistic science tells us that what we dream is merely a reaction to our daily life, as we perceive it, and that the images we see in our dreams are produced by firing brain neurons. That both of these claims are patently false is another topic, another book. What is important to us is the main difference between dreams and hallucinations: the latter are open-eyed visual perceptions of objects or entities appearing not sourced in the material world. Since, as we are told, that is not possible, all hallucinations are either dreams, pseudo-hallucinations or symptoms of some kind of physical or mental pathology.

At the bottom of the materialistic paradigm lies the postulate, or the dogma, that everything real must be sourced in that which we call matter, and hence the *objective* reality can only be perceived through carnal senses. And dogmas, by their very definition, are oblivious to counterexamples. They are also immune to influences by events with obvious alternative explanations, even when these events are experienced by the majority of the human population.

So far, we have established the ubiquity of the phenomenon of hallucinations. We will reinforce this aspect of hallucinations in the subsequent chapters. This in itself is insufficient challenge to the materialistic tenets. However, it makes the phenomenon much more normal and common than what we are conditioned to believe. The ubiquity of hallucinations also makes them more relevant to our understanding of reality — what is it that we so frequently perceive with eyes wide open?

There is *a* materialistic reality, and there is a much bigger reality, partially accessible to us via extra-sensory perception. One cannot understand the former without at least acknowledging the existence of the latter. By acknowledging the normality of the phenomenon of hallucination we are de facto accepting the thesis that the *large* reality exists in some way. This is precisely where the ubiquity of the perception of hallucinations by (normal) people leads us. With that we have tentatively made the first step in our long journey of understanding.

Chapter 2

Hallucinations: Minor or Major?

Lead me from the unreal to the real;
Lead me from darkness to light.
A prayer from *Brihadaranyaka Upanishad* — an
ancient Vedic text

2.1 Regarding Minor Hallucinations

A hallucination is minor if it can be readily explained away as a regular or normal percept. For example, you feel a tap on your shoulder, and when you turn around you see no one. There is, then, a simple explanation, the one we are conditioned to look for — it must have been a muscle twitch. Your life will hardly be affected by such a mundane episode, even if there is a trace of doubt, and for a moment you fantasize that someone invisible touched you. The muscle twitch explanation is much less *extravagant,* hence much more acceptable.

Where does mundane end, and extravagant start? The following very intriguing anecdote conveys deep wisdom that implies a rather fuzzy boundary between mundane and extraordinary. From 1974 to 1975, for about a year, the renowned parapsychologist and nuclear physicist, Edwin C. May, visited India. His goal was to research psychic phenomena brought about by yogis and other initiates. One day he was told of a wise man in a not-too-distant monastery, who could materialize objects inside intact coconuts. So, May took a coconut and walked for hours to reach the monastery, where he found the wise old man. He asked him if he really could materialize an object inside the coconut as he was holding it. The wise old man replied through a translator: "Yes, I could materialize a physical object inside your coconut by the sheer force of my spirit. But, what makes you think that you could hold the coconut?" [1] Indeed! What makes us think that we and our coconuts exist? After all, according to our own science, we are conglomerates of elementary particles of dubious physical existence.

That being said, we will continue pretending for a while that twitching muscles is a much less miraculous event than a tap on the shoulder by an entity that *clearly* does not exist. Similar conventions will allow us to tentatively recognize *minor* hallucinations.

Brief and unsubstantial hypnagogic and hypnopompic (H&H) hallucinations are very easy to explain away as dreams or sleepy-eyed misperceptions. I once *saw* — with my eyes wide open — a mandala, seemingly floating above my bed. For a few seconds I stared at it indifferently and with mild curiosity, while it stayed fixed in a vertical position. The moment I started to scrutinize the scene, trying to relate it to the *normal* reality, the mandala moved away from me, and smoothly merged with the bookshelves behind it. It was as if it was offering me the trivial interpretation: "You fool! It was the bookshelves you were observing all along!" And so, the rude mandala did not earn an entry for itself in my files.

Like H&H hallucinations, visions or hallucinations of the dying happen at states bordering the waking state of consciousness. We mentioned in the Introduction and in Chapter 1 that according to some studies this is a fairly common phenomenon, likely experienced by the majority of unsedated dying people. In most of these cases dying people perceive beautiful scenes, angelic entities or predeceased relatives and friends. For example, here is a case of a dying old woman:

> *Then she pointed up and said, "Look." We [her son and her daughter] looked and there was nothing there, so we asked what*

she was pointing at. "Can't you see them? The angels?" she said. We told her we couldn't. Then she said, "No, I suppose you can't. Just me. But they are so lovely, I wish you could see them." [...] Then she died.[2]

A rather easily explainable hallucination! Or is it?

The following case is similar. It was recorded some 80 years earlier by Sir William Barrett, and narrated by his wife, who was a gynecologist. A woman who was dying after childbirth had a similar vision:

Suddenly she looked eagerly towards one part of the room, a radiant smile illuminating her whole countenance. "Oh, lovely, lovely," she said. I asked, "What is lovely?" "What I saw," she replied in low, intense tones. "What do you see?" "Lovely brightness — wonderful beings." It is difficult to describe the sense of reality conveyed by her intense absorption of the vision. Then — seeming to focus her attention more intently on one place for a moment — she exclaimed, almost with a kind of joyous cry, "Why, it's Father! Oh, he's so glad I'm coming, he is so glad."[3]

Despite the profusion of similar cases — there are hundreds, perhaps thousands of recorded instances of *minor* visions of dying people — they are hardly paradigm changers. The simplistic medical explanation — reducing the phenomena to fantasies of sick, clouded brains — seems reasonable, if its scope is confined only to minor visions.

Subjectively, the appreciation of the power of these visions belongs almost entirely to the dying people themselves; the living, being mostly blind to the phenomenon, are stuck in a quasi-superior detached point of perspective. And why would the living act differently? Doesn't our live experience teach us that it is always other people who die?

The point we have barely started making is that the reality that we normally perceive is very, very shallow. Hallucinations (of the dying or of the living) open windows through which we can take glimpses of a much deeper and, in many ways, more valid worlds.

The book of visions of the dying would have been closed in silence if the story ended with minor hallucinations. It doesn't! It has only started! In order to indicate the direction of our narrative, here is a case that is rather similar to the above two cases, except for the finishing twist:

He was unsedated, fully conscious, and had low temperature. He was a religious person who believed in life after death. We expected him to die too, as he was asking us to pray for him. [...] Suddenly he exclaimed: "See, the angels are coming down the stairs. The glass has fallen and broken." All of us in the room looked toward the staircase where a drinking glass has been placed on one of the steps. As we looked, we saw the glass break into a thousand pieces, without any apparent cause. It did not fall; it simply exploded.[4]

Cases that have potential to become paradigm changers start with veridical hallucinations. The next chapter will be devoted to such hallucinations pertaining to (what we call) death. We also notice the asynchronicity of a single event, the breaking of the glass, from different viewpoints. In Chapter 6 we will discuss more thoroughly the phenomenon that we have barely touched with this anecdote: Time.

2.2 Minor Auditory Hallucinations

Let's now turn our attention to minor auditory hallucinations, particularly to minor auditory verbal hallucinations (AVH). Below our thematic threshold are tinnitus-like sounds,[5] while hallucinations of sounds of emotional outbursts (cry, laughter) are sporadic and rarely reported.

Hallucinations of music are not widespread, but they are virtually common among musicians.[6] This is not surprising, since musicians in general seem to be amply endowed with paranormal perceptional talents. For example, Charles Honorton's experiment on projecting images during ganzfeld (reduced external sensory) conditions, repeated jointly with Ray Hyman, a skeptic who imposed more stringent conditions, produced positive results astronomically beyond chance, especially when the subjects were musicians (25% chance, 35% general study, 50% musicians!).[7]

The interesting musical hallucinations start with auditory hallucinations delivering new music, along the way making composers out of the hallucinators.[8] An example of a notable composer who created his music through hallucinations is Robert Schumann (1810–1856). His wife, Clara, wrote that "he heard entire pieces from the beginning to end, as if played by a full orchestra".[9]

We close our very brief overview of musical hallucinations with an intriguing anecdote featuring the Italian composer Giuseppe Tartini (1692–1770). Tartini's account of his experience was published by the French astronomer, Jerome Lalande, in his travel memoirs:

One night I dreamed I had made a pact with the devil for my soul. Everything went as I desired: my new servant anticipated my every wish. I had the idea of giving him my violin to see if he might play me some pretty tunes, but imagine my astonishment when I heard a sonata so unusual and so beautiful, performed with such mastery and intelligence, on a level I had never before conceived was possible. I was so overcome that I stopped breathing and woke up gasping. Immediately I seized my violin, hoping to recall some shred of what I had just heard; but in vain. The piece I then composed is without a doubt my best, and I still call it "The Devil's Sonata," but it falls so far

short of the one that stunned me in my dream that I would have
smashed my violin and given up music forever if I could but
have possessed it.[10]

His dream might well have been a combination of auditory and visual H&H hallucination.

I have difficulty classifying any AVH (auditory verbal hallucination) as a minor experience. After all, as we have indicated in the Introduction, even a single episode of "hearing voices" could be enough for the experiencer to be diagnosed with schizophrenia. Perhaps the only way to minimize an AVH experience is to follow the algorithm we have been socially implanted with: pull the *scientific* blanket over our mind and reduce AVHs to anomalous brain excretions. This explanation is feasible up to a point but fails flatly in case of AVHs with veridical elements.

The following example is rather instructive, since the hallucinator was the neurologist and medical expert in hallucinations, Oliver Sacks. While in his kitchen one Sunday morning he heard someone knocking at the door; his friends Jim and Kathy entered through the open door, and he offered them eggs for breakfast. Jim asked for "sunny side up" and Kathy liked them "over easy." The conversation was done without Sacks seeing his friends. Five minutes later, when he entered the living room, he was flabbergasted to see that there was no-one there. His conclusion: "the whole conversation, at least their side of it, had been completely invented by my brain."[11] By faithfully sticking to the *safe*, materialistic cliché, he forfeited his chance: a door of perception, as per Aldous Huxley,[12] had been opened ajar for him, and he promptly slammed it shut.

Since AVHs are by their nature almost always powerful experiences, we will also deal with them within the context of major hallucinations.

2.3 From Minor AV Hallucinations
to Major AV Hallucinations

The following account was given by a 19-year-old woman:

> *I was thirteen when I had a Voice from God for my help and*
> *guidance. The first time that I heard this Voice, I was very much*
> *frightened [...]. I rarely hear it without its being accompanied*
> *also by a light. The light comes from the same side as the Voice.*
> *When I heard it for the third time, I recognized that it was the*
> *Voice of an Angel.*

Is this a minor, or is it a major hallucination? Despite combining AVH with visual hallucinations, and despite multiple occurrences, the fact that it was a one-person experience and the lack of veridical components render the anecdote insignificant. However, if we use the impact on our recorded linear history as the main criterion for its significance, the above AVH may be one of the most consequential ever! The experiencer was the French heroine, Joan of Arc (1412–1431), who, through her visions, profoundly influenced the last decades of the Hundred Years' War between the French and the English, contributing to the eventual French victory. The first half of the fifteenth century was the critical time when Atlantic European states positioned themselves for colonial expansion, which in turn deeply affected all of our planet's history ever since.[13] Joan herself was burned at the stake by the English. Thrice!

In 1920, Joan of Arc was canonized as a saint. Nowadays she would likely have been diagnosed with at least proto-schizophrenia. However, AVHs combining veridical elements are not that easy to wash away. Here is a one-word verbal *hallucination* of such a sort:

> *A voice shouted to him [the sculptor Jean Dupré], "Stop!" as*
> *he was driving along a deserted path of the Apennines. The*
> *voice was also heard by Mme. Dupré, seated by the side of her*

husband ... [The scene repeated three times.] The travelers, now thoroughly perturbed, all the more so as Mme. Dupré was trembling with fear at the sound of the mysterious voice, finally decided to get down. The sculptor then noticed that the axle of one of the wheels was loose, and so the carriage was in the point of rolling down a precipice skirting the road. [14]

What to make out of the fact that this was a portent heard by two people at the same time? This aspect of the anecdote only, excludes the orthodox medical explanation of the event being a simple, brain-generated auditory hallucination. In comes the "it-is-a-hoax" all-purpose apologia.

With the preceding example we started wading into the deeper waters of veridical hallucinations, from which there is no rational return to the safety of the herding corrals of materialism. We keep on going, deeper and deeper. In September of 1982, Rachel Clark was driving her van when she was "told by a voice as clear as if someone were sitting beside her to get out of the car." [15] After hearing the voice several times, she finally did as she was told, taking her three baby children with her. Moments later the car exploded into flames.

One more step: Teryn was a two-month-old baby, diagnosed by the pediatrician, Melvin Morse, to have been afflicted with an incurable Wolman's liver decease. The findings were double-checked by several specialists in a number of ways (including by electronic microscope). Morse himself cried when he informed Teryn's parents that their baby was doomed. But the baby did not die! In fact, she started thriving, and by one year she was diagnosed as cured. The specialists Morse consulted were as astonished as him. The catalyst for the miraculous cure was Teryn's uncle, whom Morse interviewed:

He described himself as not being a spiritual man and felt more comfortable at a party than in a church. Here is what he said:

"I had been at work and came in around nine at night. I went into the bedroom and lay down to go to sleep. I was almost asleep when I heard a voice say, 'Go to you niece's house and touch her abdomen.' I woke up completely and looked around. I could not really believe I had heard anything. The voice was loud and insistent, and I almost thought about getting up and doing what it told me to do, but I didn't. 'I'll do it tomorrow,' I said. I lay back down. Then the voice spoke up again. It said, 'Go tonight. Tomorrow will be too late'." He immediately got out of bed and left the house. He did not tell his wife where he was going. He just got into his car and left, arriving at Teryn's house about ten o'clock. Although he hadn't been to the house in several months, the family thought he was there to lend emotional support and did not thing it was terribly odd that he came by. He asked if he could hold Teryn and then took her to another room. Mrs. Hedlund [the mother] said, "Why are you doing this?" He replied, "I want to be alone with her please." He put his hand on her abdomen as he had been told to do by the voice. The hand became warmer and warmer. He stayed there for a few minutes and then abruptly left. He later [but before it was apparent that Teryn was recovering] only told his father-in-law.[16]

Morse summarizes: "The coincidences are just too great for me to think that the [uncle's] actions were not somehow associated with Teryn's outcome."

The above cases are but a marginal sample of the recorded instances of *hallucinated* voice-interventions and various other manifestations.

2.4 From Minor Visual Hallucinations to Major Visual Hallucinations

Here is a minor H&H case that happened to me on April 3, 2007.

I went to bed at around 11 pm, read a bit and started dozing off. Then I sensed someone's presence in the room, and I opened my eyes. I was still lying in the reading position, my head propped against the pillow, so that when I opened my eyes, I faced the opposite side of the bedroom. I saw a woman by the bookshelves, close to my feet. Naturally, I thought it was my wife I was looking at, and so I observed her calmly, expecting the sleepy-eyed fuzziness of the scene to clear up and match my expectation. I wondered what she was doing and what she was going to tell me. The woman was slightly stooped over the bookshelves, as if browsing the titles of my books. Then she turned towards me; it was not my wife! She was a stranger! She stood motionlessly, as if surprised that I was perceiving her. For a few moments we stared at each other, both of us seemingly flabbergasted. And then something happened that I surely wasn't expecting: the woman started floating towards the ceiling, and simultaneously she started dissolving into thin air. In a couple of seconds she was gone! My heart started pounding; I needed a minute to calm down!

That was my first H&H vision, and it baffled me. For the next two sleepless hours I tried to understand what had just happened. I saw a *ghost*, and ghosts, as "everyone knows," don't exist. Moreover, it appeared to me that the ghost saw me! What is it really a ghost that I saw? And was I a ghost to my ghost? I couldn't find a satisfactory explanation.

My ghost encounter was objectively a minor hallucination. Nevertheless, it was a watershed event in my life: my curiosity was unleashed, and ever since I have been trying to induce such visions. Nothing impacts one's view on the so-called paranormal phenomena more than a direct personal experience.

As we will see in Chapter 9, my experience with hallucinations eventually became comparable to the H&H visions of some other prolific hallucinators. Take, for example, Gerolamo Cardano, a.k.a. Jerome Cardan, today mostly remembered as the person who first published certain basic mathematical formulas. Here is a quote from his autobiography *The Book of my Life*,[17] (pages 147–158):

> *These images arose from the lower right-hand corner of the bed, and, moving upward in a semicircle, gently descended on the left and straightway disappeared. The images of castles, of houses, animals, of horses and riders, of plants and trees, of musical instruments, [...], of men of divers costumes [?] and varied dress; images of flute-players, with their pipes as it were, ready to play, but no voice or sound was heard. [...] I could see a veritable chaos of innumerable objects rushing dizzily along en masse, without confusion among themselves, yet with terrific speed. These images were, moreover, transparent, [...] yet so dense as to be impenetrable to the eye. [...] Even flowers of many a variety, and four-footed creatures, and divers birds appeared in my vision.*

There are thousands of "haunted houses" ghost stories in hundreds of collections. This is not a book meant for spooking readers, so we will mostly pass this subject. We will only recount the following ancient anecdote, and then slowly phase out the ghost terminology; the word "ghost" itself lost its original meaning and is now loaded with empty Hollywood-made ridicule.

Pliny the Younger lived 61–113 AD; he recorded the experience of his Greek friend, Athenodorus, the keeper of the library at Pergamon, that happened during one night in a cheaply rented room in Rome that nobody wanted because it was haunted! "He sat up all night to see the ghost and was rewarded by the apparition of a man in chains that clanked as he walked. The form beckoned and Athenodorus followed outside, to where the figure pointed at the ground. The next day, acting on Athenodorus' information, the magistrates had the spot dug up and found the skeleton of a man in chains." This story is the origin of the stereotypical "ghost in chains" image. It is meaningful to us because of the veridical ending.

The following anecdote combines interactive hypnagogic hallucination with the phenomenon of *timeslip* or retrocognition; we will cover *hallucinated* timeslips more extensively in Chapter 6. The quote below is from the book *The Ghostly Register* by Arthur Myers. The experiencer is the actress, Andrea Haring. Her acting troupe moved into The Mount, a neo-Georgian mansion built between 1900 and 1902 by the American writer, Edith Wharton.[18]

It was in the winter, about 1979. I went up to Edith Wharton's writing room because a couple of people were having a discussion in my bedroom. There was an extra mattress in Edith Wharton's room, and I thought I'd just lie down till they were done. I stoked up the wood stove in the room so that it would have stayed warm until about noon the next day. This was around midnight. I slept till about four in the morning. Then I drifted awake because it was very cold in the room. My eyes were still closed, but I was awake, and I sensed there was someone in the room. I opened my eyes and saw three figures in the room, and where the room had been bare of furniture there was now a small divan and a desk with a chair. I thought I must be dreaming, but I sort of pinched myself and slapped myself, and I thought, No, I'm awake; I must be seeing ghosts.

One was Edith Wharton, whom I recognized because I'd been reading her biography. I could see the details in her dress and in her face and the way her hair was done, even though it was dark outside. She was kind of half seated, half lying on this divan. At this little desk was a man who was writing. I could see the muttonchops and his face and his outfit. I didn't recognize

41

him, but he appeared to be gesturing to Edith. Although I couldn't hear any sound, they appeared to be talking to each other. He would stop and make a gesture like, "Oh, yes" or "Aha," and then he'd start writing again. It almost appeared as though she were dictating to him. It was interesting because this was her writing room. And there was a third figure who was standing with his arms folded, and I recognized him as Teddy Wharton, her husband, whom she eventually divorced. He was standing there with his arms crossed, looking at the two of them. I thought to myself, I wonder if I can leave. The minute that thought crossed my mind, all three of them turned and looked at me. I looked from one to the other. I have no reason for why I did this, but for some reason I just kind of smiled at them and nodded my head, kind of like, "Hi, I'm here. I see you, and I guess you see me." Edith Wharton gave me a kind of short, dignified nod. Teddy Wharton gave me a kind of brusque acknowledgment, with a nod. But I felt that that was his way, not that he was malevolent towards me or anything. And the guy at the desk, whom I didn't recognize, beamed at me and nodded his head quite vigorously. And then they all turned back to what they had been doing. I felt absolutely free to go at that point, which I did. As I closed the door, I still saw them there.[19]

This story amplifies and expands my own minor experience mentioned above. It also suggests that we are, indeed, as much *ghosts* to the *ghosts* as they are *ghosts* to us.

From a collection of Celtic lore by Evans–Wentz we have the following short anecdote.[20] The narrator is Pat Ruddy, who was 80 years old at the time of the interview.

One night, however, near midnight, [...] I set out from Bantrillick to come home, and near Ben Bulbin there was the greatest army you ever saw, five or six thousands of them in armour shining in the moonlight. A strange man rose out of the hedge and stopped me for a minute, in the middle of the road. He looked into my face, and then let me go.

Five thousand ghosts at once! Were they products of Pat's imagination? Perhaps. But, then, how to explain five thousand ghosts perceived at once by five thousand witnesses? We will see such cases later (see Chapter 5).

2.5 Jesus

My friend Mikhailo[21] was a mathematician who lived in a so-called socialist country. At the time of the event — early 1980s — he was in his twenties, and he did not know, nor cared about virtually anything that was spiritually oriented. In fact, the word "spiritual" was meaningless to him, or, at best, it was associated with young ladies who read poetry. The only religion he had some knowledge of was dialectic materialism — and even that did not interest him; he accepted it tacitly only in order to be aligned with the local social flow. Mikhailo lived alone — the times were not so bad — in a one-half-bedroom apartment of a typical working-class high-rise neighborhood. One ordinary night, after an ordinary day, he went to bed, relaxed for a while, and fell asleep. He does not know when exactly, but it must have been shortly after he fell asleep, something made him open his eyes. He was not alone — there was a figure of a man by his bed. After a few seconds of staring dumbfoundingly, his bewilderment escalated, for the man he saw was crucified on a large cross apparently erected at the foot of his bed. That much he knew — Jesus Christ! The scene was illuminated, as one would expect it to be. Mikhailo stared unbelievingly at it for — what he thought afterwards — a long time. He could not make any sense of what he was seeing. He closed his eyes and opened them again; the apparition was still there. He cocked his head, and the scene changed in three-dimensional perspective;

it was not a projection on the opposite wall. Eventually Mikhailo gathered his courage and took a step in the direction of the cross, at which point the cross, Jesus and all, dissipated into thin air.

Recalling the apparition in the morning, Mikhailo dismissed it as a hallucination of no major significance, relegated it to a seldom accessed memory cell of his brain, and continued with the business of living his years. And he kept quiet. Keeping quiet seems to be the reason that visions of religious personalities are not as often reported today as they were during "prescientific times," with a notable exception of the reign of the Inquisition. Still, there are dozens of visions of Jesus Christ that have been officially endorsed by the Catholic Church, hundreds of reported visions, with the total of the quiet, unrecorded cases likely in hundreds of thousands. My own small sample marginally supports this claim. Beside Mikhailo, I know of two other people who have seen apparitions of Jesus: a girl of 16, raised in a secular family, and a woman in her seventies who saw Jesus in broad daylight and was bewildered to realize that the people around her could not perceive Him.

In some cases, visions profoundly affect the life of the perceiver. A notable example is the experience of the French polymath, Blaise Pascal (1623–1662), in whose honor we now have a unit of pressure (physics), a triangle of binomial coefficients (math), and a programming language (computers), all named after him. During the night of November 23 1654, Pascal had a profound vision of Jesus. We do not know the details of his experience, but we do know how it affected him. Immediately after the vision Pascal wrote a note to himself that he hid by sewing it into his coat. It was discovered and published after his death, eight years later:

> Grandeur of the human soul. Righteous Father, the world has not known you, but I have known you. Joy, joy, joy, tears of joy. I have departed from him: They have forsaken me, the fount

of living water. My God, will you leave me? Let me not be separated from him forever. This is eternal life, that they know you, the one true God, and the one that you sent, Jesus Christ. Jesus Christ. Jesus Christ. I left him; I fled him, renounced, crucified. Let me never be separated from him. May I not forget your words. Amen.[22]

Pascal devoted the last eight years of his life to the philosophy of higher, metaphysical subjects.

An unlikely Pascal-like experience unfolded in India some three centuries later. Sundar Singh was born in 1889 into a Sikh family. His broad-minded mother encouraged him to read old Vedic texts — he knew *Bhagavad Gita* by heart when he was 7 — and she took him to various sadhus (holy men) and priests in search for the right way for him. He practiced Yoga, and he read holy books of many religions, but not the Bible. He considered Christianity to be a "false" religion. When he was 15 years old he burned the Bible and participated in stoning of Christian preachers. But he could not "get any [spiritual] satisfaction and peace", which brought him on the verge of committing suicide. His words: "Three days after I had burnt the Bible, I woke up about three o'clock in the morning, had my usual bath, and prayed, 'O God, if there is a God, will thou show me the right way or I will kill myself'."[23] He wanted to place his head on a railway track in front of the 5 o'clock train. Then something happened at 4:30 a.m. First he saw "great lights," making him think that something was on fire. Then he saw a man, whom he describes without any hesitation as Lord Jesus Christ: "I heard a voice saying in Hindustani, 'How long will you persecute me? I have come to save you; you were praying to know the right way. Why do you not take it?'"[24] Sundar fell on his knees and felt "wonderful Peace." And that is how it started. Sundar Singh, later in his life known as Sadhu, became a Christian missionary and preacher. He died in or around 1929.

Sadhu might have requested his vision obliquely. In the following interesting example, the Jesus-apparition was intentionally and directly invoked. One evening in 1950, when Catherine North was 14, she was telling her two little sisters nighttime stories:

> One of them asked, "If Jesus is real, why can't we see Him?" Having been well indoctrinated in the power of belief and faith for all fourteen years of my life, I replied, "We can. All we have to do is believe and concentrate real hard, and He will appear to us." So all three of us concentrated, and to my amazement the brightest white light in the form of a man appeared at the foot of the bed. We all saw him.[25]

How cute in its innocence!

The little girls stumbled on a formula well known to mystic initiates across the religious spectrum: powerful and repeated invocations give rise to supernormal effects. Here is what happened in a Nestorian monastery near Bagdad, after the monks were invoking Jesus' name for a long time:

> Sure enough Jesus came to conduct the service! If the person who appeared had been normal looking, I [Murat Yagan] would have been convinced that some one of the monks had dressed up as Jesus, but this appearance of Jesus was of a man at least 16 feet tall and 6 feet wide. I couldn't enjoy the ceremony because I fainted at the apparition.[26]

Note that the last two examples are cases of group hallucinations. The following medieval account is of a mass hallucination; we mention it here as a prelude to Chapter 5, where such cases will be covered more thoroughly. It happened in the year 1188, and the quote comes from Roger de Hoveden's *Annals*:

The Heavens opened, and in the sight of many ... a cross appeared, very long and of wonderful magnitude, and it appeared as though Jesus Christ was fastened thereon with nails, and crowned with thorns; His hands were stretched out on the cross ... and His blood was flowing down, but did not fall upon the earth. This appearance lasted from the ninth hour of the day till twilight.

Dismissible as a dark-ages fantasy? Not so simple, since such or more extraordinary mass visions are spread throughout our history, including modern times. More about this topic in Chapter 5.

In visions Jesus often comes together with his mother, as in the classic scene where the Virgin Mary holds baby Jesus in her arms. Such was the vision of Saint Ignatius of Loyola, who lived 22 October 1491 to 31 July 1556. In his youth he was a valiant career soldier, until his legs were injured by a cannonball. While in the hospital he experienced daily apparitions of a serpent-like entity. Later he saw an apparition of the Virgin Mary and baby Jesus. This vision profoundly influenced his life, and he underwent a mystic or religious conversion. Like his Buddhist soulmates, he spent years in a cave (in Catalonia), praying for seven hours per day. Ignatius of Loyola became an enlightened initiate who will be featured again in our narrative (see Chapter 8), when we cover outrageous hallucinations.

Visions of motherly, angelic women, usually identified as the Virgin Mary, have been reported often in modern times — according to Michael P. Carroll,[27] there have been 230 alleged visions of the Blessed Virgin Mary between 1928 and 1975 that have been acknowledged (but not endorsed) by the Roman Catholic Church — but they seem to have been particularly prevalent in pre-Inquisition Spain. A good source of such cases is William A. Christion Jr's *Apparitions in Late Medieval and Renaissance Spain.* We will only consider the extraordinarily bizarre case of Pedro Garcia, a child of about 10; it seems that the Virgin Mary manifests mostly to children.[28]

The year is 1399; the following is an insert from the transcript of the vision document:

> *The following Thursday Pedro [the son of Yñigo] was keeping the sheep of his father [...]. Suddenly there appeared to him a lady so resplendent that he could not look at her directly, but only in parts. She told him he should tell of his vision to the parish priests, clergy, and council of [the village] Santa Gadea. [...] Out of timidity and fear he would not be believed, Pedro did not mention the apparition to anybody. The following Sunday night, at the hour of the second cock crow (!) when Pedro was in bed, the Virgin Mary came [...], and she brought with her [...] two men dressed as monks. One of them took him by the arm and pulled him out of bed, and the other took his belt and struck him many blows and wounds.*[29]

When neighbors heard Pedro's cries, they saw the house "as bright as at noonday," and found Pedro lying on the floor in shock, with whip marks on his body. Pedro did talk after that, and, as per a request by the Virgin Mary, a monastery was eventually built at a certain spot.

The Virgin Mary with two extortion thugs brought to beat up a boy? What a bizarre apparition!

In Spain, only a handful of cases of visions were reported in the years between 1525 and 1900. Nobody was eager to submit to Inquisitorial checkup when one hundred lashes were the vitamins of the day, preventively prescribed for the "health of the soul."

2.6 Miscellaneous Hallucinations

We end this chapter with three examples of induced hallucinations. The first example is of hallucinations triggered by being at a certain place under reduced sensory conditions. The second example is strongly veridical, and as such it supports the genuineness of the thesis that hallucinations are sourced in the objective reality, to the extent that such a reality exists. The third example will orient us in the direction of what will eventually become our model, according to which *objective realities do not exist*.

There are two people I am aware of who dared to spend a night in the King's Chamber of the Great Pyramid of Giza. Napoleon did it, but he kept quiet. We only know that the morning after the experience he came out of the pyramid looking rather shaken. Many years later, during his exile on St Helena, he told a friend that there was no use of talking about it and that, "You would never believe me."[30]

The traveler, adventurer and journalist, Paul Brunton, did talk. Here is his account of his hallucinations during a night in the King's chamber:

A strange feeling that I was not alone began to creep insidiously over me. Under the cover of complete blackness, I felt that something animate and living was throbbing into existence. [...] With pliant soul I waited and wondered. Gradually these [shadows] took more definite shape, and malevolent countenances appeared suddenly quite close to my own face. [...] Monstrous elemental creations, evil horrors of the underworld, forms of grotesque, insane, uncouth and fiendish aspect gathered around me and afflicted me with unimaginable repulsion. In a few minutes I lived through something which will leave a remembered record behind for all time. [...] Never again I would repeat such an experiment.[31]

The malevolent entities then disappeared.

I became conscious of a new presence in the [King's] chamber, of someone friendly and benevolent who stood at the entrance and looked down upon me with kindly eyes. In my vision the apparitions of these two beings [there were two of them!] presented an unforgettable picture. [...] There was a light a-glimmer all around them, which in the most uncanny manner lit up the part of the room. Indeed they looked more than men, bearing the bright mien of demi-gods; for their faces were set in a unique cloistral [monastic] calm.

The benevolent apparitions then warned Brunton to turn away while there was enough time, but he persisted and answered that he "must follow this way."

What happened immediately afterwards is still not very clear to me. It was as though he had unexpectedly given me a dose of some [...] slow working anesthetic, for all my muscles became taut, after which a paralysing [sic] lethargy began to creep over my limbs. [...] I appeared next to pass into a semi-somnolent condition and a mysterious intimation of approaching death crept into my mind. It did not trouble me, however, for I had long ago liberated myself from the ancient fear of death and arrived at a philosophic acceptance of its inevitability. [...] I had the sensation of being caught in a tropical whirlwind and seemed to pass upwards through a narrow hole; then there was a momentary dread of being launched in infinite space, I leapt into the unknown – and I was **free!** *[...] I gazed down upon the deserted body of flesh and bone, which was lying prone and motionless on the stone block. [...] The arms were folded across the breast — certainly not an attitude which I could remember having assumed. [...] There lay the seemingly dead form of myself, the form from which I had withdrawn. [...] A single* *idea obtained recognition in my mind, a single realization overwhelmed me. [...] This is the state of death. Now I know that I am a soul, that I can exist apart from my body. I shall always believe that, for I have proved it!*

"I was but a phantom sojourning in space," wrote Brunton.

Brunton's adventures in Egypt were not merely accidental events, for at that time he was already an initiated mystic and an explorer of the (so-called) paranormal phenomena. A year earlier he traveled through India, where for a period of time he was a disciple of the famous Indian yogi, Maharshi.

The following case comes from the files of the psychiatrist, Stanislav Grof, a renowned expert in the effects of LSD on human consciousness. During one of Grof's LSD therapeutic sessions, Richard [Grof's patient treated for depression] had a very unusual experience involving a strange and uncanny astral realm, where he encountered, or hallucinated, various beings. Quoting Grof:

I wrote down one of the messages [that Richard received from the beings he encountered] that was very specific and could be subjected to subsequent verification. It was a request for Richard to contact a couple in the Moravian city of Kromeriz in Czechoslovakia and let them know that their son Ladislav was doing all right and was well taken care of. The message included the couple's name, street address, and telephone number. All of these data were unknown to me and the patient. [...] After some hesitation and with mixed feelings, I finally decided to do what certainly would have made me a target of my colleagues' jokes, had they found it. I went to the telephone, dialed the number in Kromeriz, and asked if I could speak with Ladislav. To my astonishment the woman on the other [side of the] line started to cry. When she calmed down she told me with a broken voice: "Our son is not with us anymore; he passed away [...] three weeks ago." [32]

Pause for a moment — excluding Donkin's hoax model,[33] what is the probability that Richard's hallucinations were not sourced in the *objective* reality? That is, what are the chances that Richard hallucinated a random telephone number and a message to an imaginary man called Ladislav that led to the eerie contact with the mother of a recently deceased young man called Ladislav? Somehow Richard's hallucinations defied the very definition of *hallucinations*.

Other cases of veridical hallucinations induced by psychedelic drugs (hallucinogens) will be interspersed throughout the subsequent chapters.

The third example comes from Ferdinand Ossendowski's travelogue *Beasts, Men and Gods*, where he narrates his escape from the Bolsheviks' pogrom of Russian intellectuals following the 1917 revolution. His long journey started in 1920, along a route through Siberia, Mongolia and China, ultimately reaching the United States. One night in Mongolia he was sharing a yurta (Mongolian tent) with many people, including a Kalmyk called Tushegoun Lama.

> He [Tushegoun Lama] stood up, pushed back the sleeves of his yellow garment, seized his knife and strode across to the shepherd [lying in another part of the yurta]. "Michik, stand up!" he ordered. When the shepherd had risen, the Lama quickly unbuttoned his coat and bared the man's chest. I could not yet understand what was his intention, when suddenly the Tushegoun with all his force struck his knife into the chest of the shepherd. The Mongol fell all covered with blood, a splash of which I noticed on the yellow silk of the Lama's coat. "What have you done?" I exclaimed. "Sh! Be still," he whispered turning to me his now quite blanched face. With a few strokes of the knife he opened the chest of the Mongol and I saw the man's lungs softly breathing and the distinct palpitations of

his heart. The Lama touched these organs with his fingers but no more blood appeared to flow and the face of the shepherd was quite calm. He was lying with his eyes closed and appeared to be in a deep quiet sleep. As the Lama began to open his abdomen, I shut my eyes in fear and horror; and when I opened them a little while later, I was still more dumbfounded at seeing the shepherd with his coat still open and his breast normal, quietly sleeping on his side and Tushegoun Lama sitting peacefully by the brazier, smoking his pipe and looking into the fire in deep thought.[34]

The Lama then denied that he had done anything. Ossendowski's companion has not perceived the event, so that Ossendowski concluded that he "had become the victim of the hypnotic power of Tushegoun Lama."

How can one man momentarily superimpose surreal scenes directly into the consciousness of another man? Perhaps this could be explained as a case of masterfully executed hypnotic induction over an inadvertently suggestible person. But then, what when the hypnotized victims are 50,000 people at once (see Chapter 5)?

Tushegoun Lama, also known as Tushe-Gun Lama, and as Dambin Jansang, had a fascinating life. Besides being a Buddhist monk, he was a fearsome, merciless warrior. During the Mongolian struggle for independence, he commanded a gang of West Mongols; during one of their expeditions his gang exterminated all Chinese and Moslems from Kobdo, the Lama himself executing ten persons. His dismaying disregard for life, his own or that of other people, reminds me of Krishna's teachings in *Bhagavad Gita*; as if his actions make conscious emphasis on the fleetingness and unimportance of human corporal existence. His death is equally dramatic: he was killed by another warrior who swallowed his heart! Or was that another of his tricks?

2.7 Reality, Step 2: Out of the Box

Around 1880, Sir William Barrett and a few other "distinguished scientists" performed an experiment with a girl who correctly guessed 9 out of 14 cards pulled out of a pack. The probability that this was a random event that happened by chance (or the p-value of this event) is $6.6*10^{-13}$, or less that 1 in one trillion, a very small number! The follow-up to this is very interesting. Sir William Barrett published the results in the *Nineteenth Century* magazine, commenting that surely nobody was going to accuse the experimenters of deliberate fraud. But in the next issue, someone did exactly that. A man called Horatio Donkin[35] asked which was most likely, a $6.6*10^{-13}$ to one chance, or deliberate collusion between the children and the distinguished scientists? (This came to be known as Donkin's Law!)

Donkin's Law is a convenient way to dismiss or ignore any results or events that do not fit the materialistic narrative. Billions of events (hallucinations included), and thousands of scientific results supporting their authenticity are passed over by faking rationality with, and hiding behind the "it's a hoax" pathetic explanation. We will resolutely not do that.

Consider, for example, Grof's experiment with LSD. The likelihood that Richard, Grof's patient, guessed by chance the names of Ladislav and his parents in Moravia, the name of the city, their address and their phone number, is infinitesimally small. Not to mention the meaningful message that was relayed through Richard by the entities he encountered while in an altered state of consciousness! Since we do not subscribe to Donkin's Law, there is no alternative but to regard (with a very large probability) the result of this experiment only, as paradigm changing. It suggests that the consciousnesses of those who have "died" still exist, and it implies that the entities whom Richard encountered while under LSD were *real*, not merely products of his imagination.

It also implies that Richard's travel during the LSD séance was a shift to another "objective" reality, an observation that we will amplify in the subsequent chapters. At the moment and in a curious way, this leads us to Napoleon and Brunton's pyramid adventures — they also travelled to other realms.

Let's go on a tangent and stay with the issue of pyramids for a paragraph. Is it possible that simply being at a certain place (the King's Chamber) within the grand pyramid affects the parameters that define *our reality* and what our consciousness perceives in a tangible way? Of course! Moreover, this categorical answer is supported by scientific experiments. Here is one cluster. Large-scale experiments were performed by a group of Russian scientists led by the mathematician, Alexander Golod. More than 20 large fiberglass pyramids were built in various places in Russia between 1990 and 2017; the largest is 44 meters high and weighs 55 tons. Apparently, objects left in the pyramid show unusual properties: say, water would not freeze at -38º Celsius. In one of the experiments a column of unknown energy was detected by military radars in the air above the pyramid, 500 meters wide and 2000 meters high. At the same time, it was observed that the ozone hole in that area repaired itself in the next two months. Mice kept in the pyramid healed faster and were much more resistant to diseases and poison. Compared to the control group, seeds kept in the pyramids had a substantially larger yield, radioactive material decayed more rapidly, and much more.[36]

Returning to Napoleon and Brunton's episode, the following question is natural: what exactly changes in our bodies or minds after consuming LSD or by being inside a pyramid? We briefly introduce String Theory, the meeting place of theoretical physics and mysticism! String Theory (from the last quarter of the twentieth century) and its variants (Superstring Theory and M-theory) postulate that the elementary units of matter (and light) are oscillating strings. Many mystics, experiencers and

The Sphinx and the Great Pyramid of Giza, both of undetermined antiquity. (Photo by Hajor, 2002.)

One of the most recent fiberglass pyramids, built in 2017; according to https://www. abo.ru, it is beneficial to human health and to the environment.

"channeled" entities claim that our perception is determined by our frequencies, or by the awareness thereof. For example, well before the emergence of String Theory, the wise entity Seth, channeled in the 1960s by Jane Roberts, discusses "spectrums of matter." We postulate that being inside a pyramid, consuming LSD, or, for that matter, dreaming or dying, changes our perceptional range of "the spectrum of matter," and shifts us to other realities. Perhaps it is true that, from the perspective of our larger existence, our primary travel happens at the level of strings, not merely at the level of what we call space. We will come back to this theme and elaborate later.

What to make out of Jesus and the Virgin Mary apparitions? The orthodoxy would like to explain them away as medical hallucinations or psychotic episodes of overzealous religious fanatics. That fails in many cases, and we have seen a few here (Mikhailo, Sadhu). The real question is not whether these are "objective" experiences, but rather why they are so ubiquitous. It seems that major events and extraordinary personalities permeate time as we know it, and, if circumstances are conducive, can be perceived in many ways by ordinary normal people.

We may not be ready yet for a 16-foot-tall Jesus! That anecdote was given only to make an introductory point: we are changing the ballgame radically, and the new rules allow extraordinary phenomena. Our main goal is to support the seemingly contradictory main postulate of the ultimate reality as much as we can; that it is the most extraordinary, and, at the same time, the simplest reality.

In this chapter we have encountered our first cases that cannot be reconciled with the materialistic paradigm. As such they are either declared as fictitious, or flawed in some way, or they're simply brushed aside as insignificant challenges to orthodox science. This will be harder and harder to accept as we keep going.

Chapter 3

Hallucinations and Dying

I can assure you that death is another beginning.
Seth, via Jane Roberts

3.1 At the Moment of Dying: External Views

William Blake (1757–1827) was a poet, a painter and a hallucinator of high order. He regularly hallucinated various sentient beings. He was, as he wrote, "under direction of messengers from heaven, daily and nightly." In a letter to a friend who lost his son, Blake wrote: "I know that our deceased friends are more really with us than when they were apparent to it mortal part. I lost a brother, and with his spirit I converse daily ..."

He claimed that his poetry was dictated to him by authors who were in "eternity." Apparently, these authors were very talented, as illustrated by the following beautiful verse:

> To see a World in a Grain of Sand
> And a Heaven in a Wild Flower,
> Hold Infinity in the palm of your hand
> And Eternity in an hour.[1]

Blake's visual artwork was also inspired by his hallucinations. In fact, he often depicted what he had *seen*, as was the case with the following sketch. At times, as we are told, he "sketched his spiritual visitants as if they were posing. If the vision disappeared, he stopped working until it returned."[2]

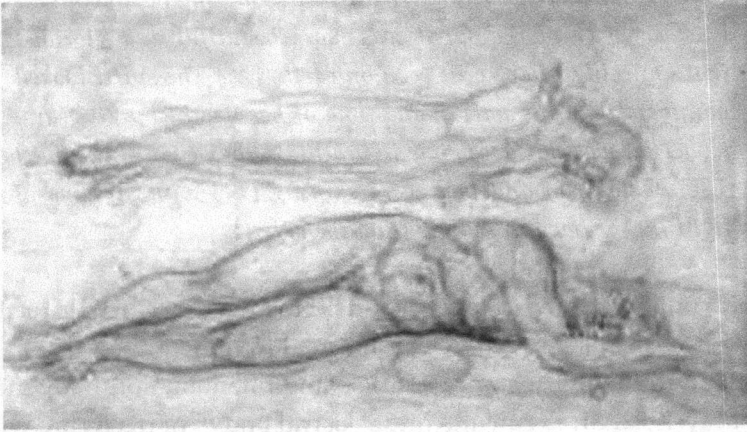

The Soul Hovering over the Body Reluctantly Parting with Life by William Blake, c.1805.

What Blake called "soul" is sometimes referred to as an astral double, spiritual body or variants thereof. Ancient Egyptians had a short term: *ka*. *Ka* is a kind of conscious energy, a spiritual double released by the dying body. After evolving out of the shape-constraints, *ka* becomes *akh*. Both *ka* and *akh* have been seen by many people throughout recorded history. In this section we will recount some such episodes of waking hallucinations of astral doubles/souls/spirits/kas, most of the time perceived by unsuspecting, normal people during other people's last moments of life.

Let's start with a simple, yet archetypal anecdote.[3] It happened in El Paso, Texas, 1907, and the experiencer was an "unnamed but reliable businessman and a newspaper editor." He was bending over his wife when she expired. Moments after, he perceived an opaque "mist, like a white fog" that rolled away from his wife's body, slowly coagulating into a full holographic image of her.

*I saw this spirit clearly [...]; there was no mistake about it.
[...] The face of the spirit was more beautiful and glorious
that anything I have ever seen on earth. [...] The spirit looked
younger [than his wife at the time of her death] by 20 years.
[...] That was the face of one in health, in the prime of life. [...]
The whole form and face were shining [...] with streams of
light that seemed to radiate from the spirit.*

The "image" lasted a few seconds. However, according to the
narrator, it was perfectly clear and real to him.

The following instance of the same phenomenon happened
in May 1902. Mr G saw the nude "astral body" of his dying wife
floating above her physical body. This very interesting account
deserves a longer quote.

*I happened to look towards the door, when I saw, floating
through the doorway three separate and distinct clouds of
strata. Each cloud appeared to be about four feet in length,
from six to eight inches in width, the lower one about two feet
from the ground [...]. My first thought was that some of our
friends were standing outside the bedroom smoking, and that
the smoke from their cigars was being wafted into the room.
With this idea I started up to rebuke them when lo! I discovered
there was no one by the door, no one in the hall-way, no one
in the adjoining room. Overcome with astonishment I watched
the clouds, and slowly, but surely these clouds approached the
bed until they completely enveloped it. Then, gazing through
the mist, I beheld standing at the head of my dying wife a
woman's figure about three feet in height, transparent, yet like
a sheen of brightest gold, a figure so glorious in its appearance
that no words can be used fitly to describe it. [...] In all of
its splendour and beauty the figure remained motionless with
hands uplifted towards my wife [...].*

At times the suspended figure would lie perfectly still, at other times it would shrink in size until it was no larger than perhaps eighteen inches, but always was the figure perfect and distinct. [...] When the body diminished in size, it struggled violently [...] in an apparent effort to escape [from the physical body, with which she was attached by the ubiquitous silver cord]. [...] Interruptions, speaking to my friends, closing my eyes, turning my head, failed to destroy the illusion. [The struggle lasted for five hours.] [...] With her last breath and last gasp, as the soul left the body, the cord was severed suddenly and the astral figure vanished.[4]

Note the perceived presence of 'delivery angels' at the beginning of the quote; we will see other similar cases further on.

Here is a more recent example recounted in detail by Dr R.B. Hout:

[My vision] was called to something immediately above the physical body, suspended in the atmosphere about two feet above the bed. At first I could distinguish nothing more than a vague outline of a hazy foglike substance. [...] But as I looked, very gradually there grew in to my sight a denser, more solid condensation of this inexplicable vapor. Then I was astonished to see a definite outline presenting themselves, and soon I saw the foglike substance was assuming a human form.

Soon I knew that the body I was seeing resembled that of the physical body of my aunt; the astral body hung suspended horizontally few feet above the physical counterpart. I continued to watch and the Spirit Body now seemed complete to my sight. [...] The eyes were closed as though in tranquil sleep, and a luminosity seemed to radiate from the Spirit Body.

As I watched the suspended Spirit Body, my attention was called to a silver-like substance that was streaming from the head of the physical body to the head of the spirit double. Then I saw the connection-cord between the two bodies. As I watched, the thought, "The silver cord!" kept running through my mind, I knew, for the first time, the meaning of it. This "silver cord" was the connecting link between the physical and the spirit bodies even as the umbilical cord unites the child to its mother.

The cord was attached at the occipital protuberance immediately at the base of the skull. Just where it met the physical body, fanlike, and numerous little strands separated and were attached separately to the skull base. But other than at the attachments, the cord was round, being perhaps an inch in diameter. The color was a translucent luminous silver radiance.

The cord seemed alive and vibrant energy. I could see the pulsations of light stream along the course of it, from the direction of the physical body to the spirit double. With each pulsation the spirit body became more alive and denser, whereas the physical body became quieter and more nearly lifeless.

By this time the features were very distinct. The life was all in the astral body; the pulsations of the cord had stopped.

I looked at the various strands of the cord as they spread out, fanlike at the base of the skull. Each strand snapped, the final severance was the hand. A twin process of death and birth was about to ensue. The last connecting strand of the silver cord snapped and the spirit body was free.

Then came the dramatic moment when the luminous body rose up from its reclining position. The closed eyes opened and a smile broke from the radiant features. She gave me a smile of farewell, then vanished from my sight.[5]

An otherworldly farewell smile! Interestingly, it is not a unique occurrence. About half a century earlier, E.W. Oaten published the following case — he was at the deathbed of his girlfriend:

A smoke-like vapour rose from the dying body and stayed at a few feet above. Gradually it became "an exact duplicate" of the girl. The "duplicate" was united to the corpse by "an umbilical cord." This eventually snapped. [...] The floating form assumed an upright position. She turned to me, smiled, and floated away.[6]

From these four examples, as well as from the cases that will follow, one gets a very strong impression of consistency — as if they all describe Blake's sketch. Unless it is the case of coincidental imagined scenes — which is hugely unlikely — the obvious explanation is that the perceived phenomena are in some way *objective*.

Sometimes there are transcendental entities who assist in the release of the spiritual double from earthly bonds. The next narrator is Lily M. Price from Australia:

During June, 1938, about 2 a. m., I was watching by the bedside of a dying girl [...]. I saw the child's [discarnate] grandmother appear at the head of the bed. She was as clearly defined and as real looking as in this life. She stayed near the head of the bed and began to move her hands as if she were trying to gently draw something outwards and upwards from the child's head. Then I saw a mist rise from the girl's head. The mist gradually took the shape of the child's bodily form. All the while the grandmother's hands were moving in a cupping fashion — just as a nurse might try to help a physical child to be born. I saw the child's form drawn out to only a little below the waist, and then the vision faded.[7]

Delivering a child at death rather than birth! The difference may not be as substantial as most of us believe. I am, of course, not saying something very original. About 2.5 millennia ago the Greek tragedian Euripides [480–406 BC] wrote: "Who knows if this experience that we call dying is not really living, and if living is not really dying?"

The following beautiful — to the extent that we can use that attribute when referring to dying — anecdote backs up Euripides' statement. It happened in 1939 when a certain Dr Hanner was a young general practitioner in rural Michigan. He was called to attend a dying 7–year–old girl:

> *Later one afternoon in August [1939], just as Hanner had finished examining the [...] girl, he heard a rustling sound off to his right side. "I assumed that I was hearing the sound of the mother's dress, so without turning around I expressed my sad regret that I didn't think there was anything more that I could do for little Mary." [...] When there was no response, Hanner turned to see a beautiful young woman [...] standing just a few inches from him. "She had dark brown shoulder-length [hair], and she was dressed completely in white," he recalled. "I knew she was not any member of the family, for the oldest of the [...] children was only 13. She seemed not to pay attention to me, and she brushed by me to approach the girl in her bed." Hanner was astonished when the lovely young woman bent over the child, then lifted the spirit from the little girl into her arms. "I was so stunned by the action that I thought my knees would buckle. [...] Little Mary's spirit was identical to her physical body, except it was translucent, and it no longer bore the ravages of the terrible illness that she had suffered for over a month." [...] The beautiful lady in white passed right through the wall with the child's spirit in her arms.*[8]

Hanner went to tell the family of the death of Mary; he also told them of the beautiful young woman, hoping to alleviate their grief. Then he happened to glance at a photo of a young woman, and he immediately recognized the *angel*. He was told that she was the sister of the father, who died at 18, killed in a car accident.

There are scientific studies supporting some aspects of *objectivity* of the phenomena we have illustrated in this section. Quoting from a 1987 study by Janusz Slawinski:

> *Recent research into spontaneous radiations from living systems suggest a scientific foundation of the ancient association between light and life, and [the] biophysical hypothesis [that] the conscious self [...] could survive death of the body. All living organisms emit low-intensity light; at the time of death, that radiation is ten to 1000 times stronger than that emitted under normal conditions.*[9]

The article was controversial at the time, as it is the case with any heresy that challenges the axiomatic dogmas of materialism.

3.2 Perceiving Distant Dying

Perceiving phantoms, apparitions, or ghosts of the dying or dead is a fairly common occurrence, and its ubiquity has been noted by many chroniclers throughout our recorded history. For instance, around 1640 Henry More wrote the following in his treatise *The Immortality of the Soul* (1642):

> *The examples [...] of the appearing of the Ghosts of men after death are so numerous and frequent in all mens [men's] mouths, that it may seem superfluous to particularize in any.*

Indeed, the following event happened about a century earlier:

> *The very day my father died [...], as I had heard nothing of his illness, still less of his death, being in a garden with my companion at about eleven o'clock that night, I began to shake*

the fruit from a pear tree. No sooner had I left the spot than I saw in front of me the form of my own father, all white, though of superhuman size [sic!]. He advanced to embrace me, but I cried out so loudly that my companion came suddenly running up, and the vision [disappeared].[10]

More than four centuries later, Hornell Hart, whose investigations of apparitions were mentioned in Chapter 1, echoed More's lament:

Apparitions of the living, dying, and dead are seen so often and with such conformity in essential features that the phenomenon must be accepted as a part of our reality.

This claim is also supported by statistical evidence. For example, according to a poll done in 1987, 42% of American adults believed that they have been in contact with someone who had died, and 67% of all widows believed they have had a similar experience.[11]

We will now choose cases where these apparitions of the dying or dead are perceived by one or more geographically distant people, who were not aware that the dying was imminent, and who perceived either the scene of the dying or an apparition at about the time the death occurred. Such cases are veridical and difficult to dismiss as coincidental medical hallucinations.

We start chronologically. The following event happened in the year 540, in the Abbey of Monte Cassino, Italy. The witness was Saint Benedict of Nursia (480–547), the founder of the Benedictine Order, and the person who established the Abbey in 520. The original source is *Dialogues of Gregory the Great*, Book II (594).

The man of God, Benedict, [...] rose early up before the time of matins [...] and came to the window of his chamber, where he offered up his prayers to almighty God. Standing there, all of a sudden in the dead of the night, as he looked forth, he saw a light, which banished away the darkness [...]. [...] and while the venerable father stood attentively beholding the brightness of that glittering light, he saw the soul of Germanus, Bishop of Capua, in a fiery globe to be carried by Angels to heaven. [The last moments of the dying light were seen by one other monk. Saint Benedict then ...] commanded [a man to be] dispatched that night to the city of Capua [some 100km south], to learn what has become of Germanus [...]: which being done, the messenger found that [Germanus] had departed this life [...] at the very instant [Benedict] beheld him ascending up to heaven.

We fast forward to the year 1709. The dying scene was again perceived through a window. The perceiver was a native Canadian, who, in the colonial jargon of the time, was referred to as "a savage." He reported what he was *seeing* with empathy, but without aplomb or religious connotations. The original source is a letter by Princess Palatine, dated March 2, 1719:

About ten years ago a French gentleman [...] brought with him in France a savage from Canada. One day when they were at table the savage, his face convulsed, began to weep. [After some prodding he said:] "I saw from the window that your brother had been murdered in such a spot in Canada." [The French gentleman, certain Longueil, told the "savage" that he had gone mad. The "savage" asked him to write down what he said, which Longueil did.] [Six] months later, when the ships arrived from Canada, he learned that his brother's death had occurred at the exact moment and in the very spot where the savage had seen it in a vision through the window. This is an utterly true tale.[12]

We ask the rhetorical question: how could the perception of the "savage" be explained away as a random medical hallucination? And, in the context of thousands of *recorded* similar veridical cases, how could it be explained away as being apocryphal?

The cases of perception, mostly visual, of the last moments of life of a person very far from the percipient are not as rare as we might think. Here is one more:

> *Monsieur Brusa [...] was dining in Superga near Turin. [...]*
> *Suddenly he stopped eating and began to weep, asserting that*
> *he saw his mother dying at Asti, without his having been*
> *informed of her illness in any way.*[13]

His mother, as it transpired, had indeed died at that very moment.

In the following case the apparition of a close relative at the time of her unexpected and distant dying was perceived *independently* by three people!

On the night of January 10, 1879, Charles Tweedale saw a face emerging from the panels of the cupboard and, as it became clearer, he recognized his grandmother. At breakfast with his parents, he began telling the experience when his father sprung from the seat and left the room leaving the food untouched. His mother explained after he left: "Well Charles, it is the strangest thing I have ever heard of, for when I awoke this morning your father informed me that he was awakened in the night and saw his mother standing by his bedside and that when he raised himself to speak to her she glided away."[14] The telegram informing them of the death of Charles' grandmother arrived later that day. The matter did not end up there: "... my father was afterwards informed by his sister that she had also seen the apparition of her mother standing at the foot of her bed."

The probability that three people randomly *hallucinated* apparitions of the same dying person, independent from each other and at about the same time the dying happened, should be close to zero. The probability that *all* 100 similar anecdotes given in Alex Baird's *Casebook for Survival* happened by chance must be *very* close to zero! Just Baird's study suffices to prove that the materialistic ashes-to-ashes model of the phenomenon of dying is way off the mark.

Sometimes the dead simply announce their passing to an unsuspecting relative or friend. For example, the French mathematician, André Bloch, reported the experience of his mother to the astronomer and great collector of paranormal anecdotes, Camille Flammarion.[15] One day, when she was dressing, she had suddenly seen her nephew, Rene Kraemer, who looked at her and said, as if laughing at her surprise, "Yes, indeed, I am quite dead."

Here is a similar case, also from the files of Camille Flammarion:

> On December 4, 1884, at half-past three in the morning, I being perfectly awake, rose and got up. I then had the most distinct vision of the apparition of my brother Joseph Bonnet... [...] My brother kissed me on the forehead. I felt a cold shudder pass through me, and he said, very distinctively, "Good-bye Angele, I am dead."[16]

The following narrative comes from the book *There is no Death* (1917) by the English clairvoyant and writer, Florence Marryat.[17] Her father, Captain Marryat, was on his ship during the Burmese wars, while his scurvy-stricken men were ordered on land to get some vegetables and fruit.

> As my father was lying in his berth one night, anchored off the island, with the brilliant tropical moonlight making everything as bright as day, he saw the door of his cabin open, and his brother Samuel entered and walked quietly up to his side. He

looked just the same as when they had parted, and uttered in
a perfectly distinct voice, "Fred! I have come to tell you that I
am dead."

Then Captain Marryat immediately wrote down the details in his log. It transpired, of course, that his brother Samuel died at the time of the apparition.

The laconic "I am dead," as bizarre a statement as it is, is the ultimate confirmation of passing over.

In some of the instances the recently deceased make rather nonchalant appearances. The following case is from the files of the Society of Psychical Research. It happened in December 1919; Lieutenant Larkin reporting:

The door opened with the usual noise and clatter. [...] I heard
his [fellow Lieutenant McConnel's] voice, "Hello boy!" [...]
and I saw him standing on the doorway, half in half out of the
room, the door knob in his hand. [...] I remarked, "Hello. Back
already?" He replied, "Yes. Got here alright, had a good trip."
I was looking at him the whole time he was speaking. He said,
"Well, cheerio!" closed the door noisily and went out.[18]

McConnel was killed in action that day; the above "vision" happened 5 minutes later!

We end this section with two cases of hypnagogic hallucinations, the first one containing numerous veridical elements.

The original source of the first case is the (London) Society of Psychical Research (reproduced in *Borderland* by W.T. Stead, 1891[19]). The narrator is Colonel H., who submitted his report in 1886. During the night of January 29, 1881, Colonel H., who was in London, woke up and saw his friend Major Poole, who was at the time in South Africa. Colonel H., being not fully awake, thought he was back in the barracks, and the following conversation ensued:

> I said, "Hello Poole! Am I late for the parade?" Poole looked at me steadily and replied: "I'm shot." "Shot," I exclaimed, "Good God! How and where?" "Through the lungs," replied Poole, and as he spoke his right hand moved slowly up the breast until the fingers rested upon the right lung.

Poole talked a bit more and then vanished. The case is interesting because of many reasons. Firstly, Poole was indeed killed in action at the time of the apparition. Secondly the Colonel saw Poole bearded, which he had never been during their 23-year-long friendship; he was bearded when he was killed. Thirdly, he had indeed been shot through his lungs. Fourthly, the Colonel was surprised by the uniform Poole was wearing during the vision; he did not know that not long before the British army uniforms had been changed from red to khaki.

In some of the above examples, the narrators were fully awake while perceiving their visions. So, even if we ignore the veridical elements, their cases are difficult to categorize as dismissible hallucinations. Here is one more such an example; it comes from Myer's *Human Personality*.[20]

The night after a brewer called Wünscher died, he appeared to his friend Oberamtmann, in the neighboring village:

> I knew nothing of his illness nor of his death. [...] In my dream I heard the diseased call me [...]. This awoke me [...]. Still thinking about it, I hear Wünscher's voice scolding outside [...]. Suddenly he comes into the room from behind the linen press; wildly gesticulating with his arms all the time [...], he called out, "What do you see to this, Herr Oberamtmann? This afternoon at 5 o'clock I have died." [A long conversation

ensued.] I asked myself: is this a hallucination? [sic] Is my mind in full possession of its faculties? Yes, there is the light, there the jug, there is the mirror, this is the brewer; and I came to the conclusion: [...] [I] am awake. [I] said to Wünscher: "if this be true, that you have died, I am sincerely sorry for it; I will look after your children." Wünscher steps towards me, stretched out his arms [...] as though he would embrace me. [...] [I] lifted my right arm to ward him off, but before my arm reached him the apparition had vanished.

This was not a typical hypnagogic hallucination; it is more fitting to describe it as a witness account.

3.3 Timothy Wyllie and Me: A Postmortem Pact?

Even before I contacted him, I knew a few things about Timothy Wyllie from his books — he was a traveler, adventurer, and inventor as well as a talented musician, painter and writer. During various periods of his exuberant life, he communicated with dolphins and aliens. He, like Blake, had regular contacts with "beings from eternity," that he called fallen angels. He knew life, and — having survived two near-death experiences and one spontaneous suicide attempt — he knew death. Who would not want to chat with a guy like him?

As it turned out, he was a good man too. I emailed him late winter 2013, and he promptly replied back. "Regard our life paths to have been electronically crossed and let destiny weave its wonders," wrote he in his first email to me. Thus, we started a meaningful correspondence that lasted some 20,000 words.

Throughout our contact my identity did not go much beyond my google address; I wanted to use my anonymity in an experiment. Here is an insert from our email communication, where I proposed an experiment:

DT (April 1, 2013):

... *This experiment would happen only if a couple of preconditions are fulfilled. The first precondition is that you predecease me; the second precondition is that, once you are in the postmortem state, you would have nothing smarter to do for a few moments other than to try to manifest to me, an unknown email-personality. Then one can say that the experiment has been attempted. If I am not too dense (both meanings) and actually perceive you, then that would be something. Postmortem manifestations had been recorded probably thousands of times; but I have not encountered a single one where the manifestation has been directed towards a person not known to the deceased, as a part of an experiment known to both.*

TW (April 2, 2013):

... *Gotcha. You might have to wait a while for me to die — but I do have to warn you that the last time I died, the idea of visiting a living person would have been the last thing on my mind. It's all so stunningly beautiful and fascinating that the Afterlife entirely occupies the consciousness. Also, I suspect it's bonds of affection which sometimes allows the dead to visit the living and even then, as we know, it's pretty unusual. You are probably better off starting with trying to contact Granny.*

Given these elements I don't hold out much hope. You might have more success reversing it and visiting me after you die. As my assignment here has no current sell-by date I'll probably still be around.

I realized that I was being presumptuous in tacitly assuming our order of departure, and so I dropped the subject. In any case I took it that the proposed experiment was rejected, and it gradually evaporated from my conscious memory.

On October 4, 2017, more than four years after our contact ended, Timothy Wyllie died. On October 17, 2017, when I was still unaware of his passing, he visited me as a *hypnagogic hallucination*. Am I certain it was him who manifested? Most emphatically, no! At the time I did not even relate the apparition to him; I merely recorded it in my files. Months later, after I learned of Timothy's departure, I began to connect the dots. Now I *believe* it was him. Unfortunately, and due entirely to the limited range of the spectrum of my perception, one dot was missing in order to raise this belief to the level of a personal proof!

That night he appeared much younger than the 77 years that he had at the time of his death. Sporting long hair and wearing blue jeans, he came about as a hippie from 1960s, which is what he looked like almost his entire adult life. He appeared briefly but rather clearly three times. Each time the apparition pointed at something, beckoning me almost frantically to look at that direction. But I could not see what he was pointing at; I was too fascinated by the sight of a hippie in my bedroom to look in other directions. In all three instances the visions lasted a few seconds, then vanished almost instantly.

Was he trying to show me a validation of our postmortem experiment? Was he pointing at one of his books? Or was it *Urantia*, the huge, miraculously delivered book, and Timothy's "most personally significant book?" Whatever it was, by failing to perceive it, I think I have lost my experimental proof of the veracity of postmortem contacts, and of their independence of personal bonds, or even of acquaintanceship.

3.4 Apparitions of the Dying: Postmortem Pacts

Fortunately, there exist many other similar cases of prearranged postmortem manifestations where the apparitions were clearly perceived. The source of the following story is S. Ralph Harlow's *A Life After Death*.

[Anna, Harlow's sister] was awaken one night by a touch on her hand. [Her husband] was away [...] and Anna was alone in the house except for the children. Sleepily she groped for a child who had padded out of her room to join her mother, but felt nothing. Then, fully awake, she looked up to see her sister-in-law, Marguerite [...] standing beside her bed. Then the vision vanished and she was alone in the room. The next morning she received a telegram from [her husband, informing her of the death of his sister and his intention to go to the funeral.] It was then that Anna remembered a pact that the six of us had made some years before ... [...] We had all agreed that upon our deaths we would attempt to communicate with those of us who were still alive.[21]

There is a short sequel to this:

That afternoon, with the children all at school and the house quiet and empty, Anna strolled in the garden. [She looked up at her second-story bedroom window.] There, holding the white curtains aside, was Marguerite waving at Anna but remaining silent. Marguerite left the window and then reappeared, waving again, and three times this silent sequence was repeated.

A triple manifestation again!

By the way, Harlow starts his book with the following eloquent soliloquy: "Who can reduce to a formula, or duplicate in a laboratory, the wonderful and beautiful apparition of my wife that I saw one May morning ..."

Another dying pact, recounted in *Dissertations upon Apparitions of Angels, Demons, and Spirits, and upon Ghosts* by Don Calmet, 1746:

Two noblemen [Marquis of Rambouillet and Marquis of Précy] were discussing the after life, as men who were not entirely *convinced of all that is said of it. They promised each other that the first of the two who should die would appear and bring news of the death to the other. [...] Six weeks afterward, [Marquis of Précy] heard the curtains of his bed being drawn. Turning to see who it was, he perceived the Marquis of Rambouillet in a half-jacket and boots. He rose from his bed to embrace him, but Rambouillet, stepping back several paces, told him that he had come to fulfill his promise that all that was said of the other life was true; that he (Précy) [...] would soon die. [...] Then Rambouillet seeing that he did not believe what he had said, showed him the spot where he had received a musket wound in the back, from which the blood still seemed to be flowing.*[22]*

The Marquis of Précy received soon afterward, by letter, confirmation of the Marquis of Rambouillet's death; he was killed in a battle. A few days later, as predicted by the apparition, the Marquis of Précy died too.

We note in passing that most of the episodes recounted in this chapter contain at least one element of veridicality — the perceivers were not aware of the death of the person who manifested.

Another sentimental goodbye kiss case is given in Flammarion's book *Death and its Mystery*.[23] It is narrated by Angele Ximenez, a lady whom Flammarion knew for many years; he vouched for the authenticity of the event based on the character of that lady. When Angele was a young girl of 16 she had a very good friend of her age, Jeanne, who one day made her promise that whoever dies first would come and say goodbye to the other, and would kiss her one more time. About six months later, three days after having a good time during Jeanne's birthday party, Angele had a vision:

Towards midnight I awakened, uttering cries of terror. Jeanne was there before me. My grandmother got up, and tried to calm me, but nothing could prevent me seeing Jeanne: she was there, and she said to me: "Good-bye! I'm dying, and I kept my promise."[...] Towards four o'clock in the morning I awakened once more; I felt Jeanne kissing my forehead. She was icy cold, and a second time she told me: "Good-bye! I'm dying." [...] My dream was real: my poor friend had died at four o'clock in the morning, the time at which she kissed me and I had felt her.

Flammarion writes (page 144) that such "observations are more numerous than one imagines."

At times the dead have difficulties fulfilling their promise to manifest to the living — helpers intervene. Two friends made an arrangement that whoever dies first the other would appear to the surviving friend in a fortnight and shake his hand. However, when one of the two friends did die, he failed to do as they had agreed upon.

Three weeks have passed and not the slightest thing had happened. Although I was glad that my dead friend had not come to shake hands, I begun to doubt there was such a thing as eternity. Besides, I soon forgot the plan. [One Sunday during early morning] I suddenly saw a completely unknown man, standing at the foot of my bed. He looked steadily at me with a smile. I was terrified and was wide awake at once. But the man went on standing in the same place and smiling at me. The most striking thing about him was his fresh complexion and happy eyes. Seeing the kindness in his eyes, I stopped being afraid and was completely master of myself again. Suddenly the man came straight up to me. If he had been a creature of flesh and blood he would, moving in that direction, have had to climb over the bed. But the man approached me as if the bed was not there, and without a word to me, passed by on my right, then through the wall and disappeared. When I turned back to the foot of my bed, I saw, to my great astonishment, my dead friend standing there. He did not smile, but gazed at me for a long time, then went the same way through the wall and disappeared. I know now that there is no death for us human beings.[24]

How easy it is to dismiss an account like this as a haphazard hallucination! How difficult it is to dismiss all such accounts as hallucinations!

3.5 At the Moment of Dying: Internal Views and Veridical Cases

Quoting Wikipedia:

When the body is injured, or if the heart stops, even if only for a short period, the brain is deprived of oxygen. A short period of cerebral hypoxia can result in the impairment of neuronal function. It is theorized that this neuronal impairment accounts for deathbed visions.

This "theory" makes as much sense as the claim that a tap accounts for tap water! To put it bluntly, it is an utterly unsustainable, false theory. To prove a theory is false, one needs a single counterexample; there are hundreds! The critique that the evidence we will exhibit is mostly anecdotal is frivolous; after all, a large chunk of medical science is also mostly based on statistically processed anecdotal evidence.

We show in this section that explaining away all deathbed visions as medical hallucinations caused by cerebral hypoxia, confusion, delirium, body systems failures, a mental reaction to stress and what not, is bogus. A much better theory explains the phenomenon in terms of shifts in the ranges of perception that happen during altered states of consciousness, from dreaming states to dying states. It is the perceived reality that changes, not the reality of perception. This is an uncontestable statement.

Its only weakness is the hugeness of the notion of *reality*. That is where we are heading! As indicated several times earlier, we are after the big fish — we want a reasonable theory of *reality* that accounts for all authentic phenomena, not just those sanctioned by the materialistic religion.

This will be done in a few chapters. Here, in this section, we will pay attention to a selected sample of veridical deathbed hallucinations. In each case dying visions correspond to supernaturally perceived events not normally known to the dying.

We start with an early medieval record found in *Dialogues* by Saint Gregory the Great, around 594 AD. When the monk Eleutherius died, his last words were "Ursus, come!" The monks were puzzled. Four days later some of the monks went to a fairly distant monastery, where they learned that exactly the same time when Eleutherius died, a monk called Ursus died too.

The next case is from *The After Death Experience* by Ian Wilson. A baby died two days after her birth. At about the same time, her great-grandmother was dying:

> Then ... she became calm and happy. It was alright, she announced she "knew what it was all about now." She very contentedly told her son that she had seen his father, her husband John, who had died in 1942 [this is happening in 1968]. Then, with a puzzled expression, she remarked that the only thing she could not understand was that John had a baby with him. She said about this, very emphatically: "It's one of our family. It's Janet's baby. Poor Janet. Never mind, she'll get over it."[25]

The case of a woman dying at childbirth and narrated by the gynecologist Lady Barrett (Chapter 2, Section 1) has a continuation; the narrator this time is the matron of the hospital who stayed with the dying woman, Mrs B., after Lady Barrett left.

I was present shortly before the death of Mrs. B., together with her husband and her mother. Her husband was leaning over her and speaking to her when, pushing him aside she said, "Oh, don't hide it; it's so beautiful." Then turning away from him towards me, I being on the other side of the bed, Mrs. B. said, "Oh, why there's Vida," referring to a sister of whose death three weeks previously she had not been told. Afterwards the mother, who was present at the time, told me, as I have said, that Vida as the name of a dead sister of Mrs. B.'s, of whose illness and death she was quite ignorant, as they carefully kept this news from Mrs. B. owing to her serious illness.[26]

The above account by the matron of the hospital was corroborated (at Sir William Barrett's request) by Mary C. Clark, Mrs. B.'s mother.

Excluding hoax as the dumbest explanation, this case alone is good enough to rule out almost all traditional models. The following cases exclude telepathy too.

A story told by Hensleigh Wedgwood, early eighteenth century:

A young girl, a near connection of mine, was dying of consumption. She had lain for some days in poor condition taking no notice of anything, when she opened her eyes and, looking upwards, said slowly, "Susan — and Jane — and Ellen!" She was recognizing the presence of her three sisters, who had previously died of the same disease. Then, after a short pause, she said, "And Edward, too!" She was naming a brother who was supposed to be alive and well in India, as if surprised at seeing him in their company. She said no more and died shortly afterwards and letters came from India two weeks later, announcing that Edward had an accident and died.[27]

From Dr Moody's book *Glimpses of Eternity* we find the story of a Mr Sykes. Sykes was dying of Alzheimer's disease and lay for months in a vegetative state, not recognizing his wife and children and not being able to speak coherently. The day of his death, the hospice workers (one of whom communicated the case to Moody) saw him sitting up in bed and talking clearly to someone he addressed as Hugh, laughing from time to time. The workers suspected that Hugh was some deceased relative of Mr Sykes, but when they told his wife she informed them that Hugh was his living brother. The next day his wife learned that Hugh died of cardiac arrest at the time Mr Sykes miraculously came to life for a few minutes![28]

In a newspaper article cited on page 180 of Janis Amatuzio's book *Forever Ours* (2004), Dr Elisabeth Kübler-Ross gave an example that "proves afterlife." A young woman is badly hurt in a traffic accident and a trucker stops by to comfort her. With her last breath the woman asks the trucker to contact her mother and to tell her that everything was all right because "Dad is already here." The good trucker travels 1000 kilometers to reach the mother (who was inaccessible by phone), and was told that her husband, the young woman's father, had died of a heart attack only an hour before his daughter expired.

We described 6 white crows, proving that not all crows are black. Are there other such cases? Of course! Flammarion: "How can we doubt when we see these manifestations pile up before us by the hundreds? No physical or historical science is founded on more numerous concordant observations."[29]

3.6 Reality, Step 3: Unextinguishable Consciousness

In Chapter 1 we mentioned the survey conducted by Karlis Osis: 5000 nurses and 5000 doctors from the USA and India were asked 11 questions pertaining to deathbed visions. From the reported sample of 3500 dying patients who were fully conscious during the last hour of their lives, 40% reported hallucinations of various personalities. It is reasonable to postulate that the majority of the remaining 60% also experienced visions of personalities but did not report them to the likely skeptical doctors or nurses. It is even more likely that the majority of fully conscious dying people experienced (reported or unreported) visions of personalities.

Even if that was all the data that we had, the Wikipedia "the brain is deprived of oxygen" explanation is hardly applicable, for it doesn't account for the thematic consistency of the reports — dying people report seeing human or humanoid entities, not random two-headed monsters or polka-dotted elephants. And, of course, "the scientific Wikipedia" ignores the rest of the "unscientific" data. Osis' report itself contains four descriptions of collective hallucinations. In one case the patient and the nurse simultaneously saw a hallucination of the dead sister of the patient. Collective hallucinations are especially damning to the materialistic paradigm, and we will devote one whole chapter to them (see Chapter 5).

The ubiquity and the consistency in the narratives of encountering disembodied entities may perhaps be explained away as a general human perceptional deficiency. The hundreds of *published* veridical cases cannot! How to explain thousands of cases of unsuspecting healthy people perceiving and interacting with holographic images of people dying far away at the exact time of their manifestation? And how to explain the many cases of dying people encountering "dead" people when, according to the information they had, these dead people were supposed to be alive?

There is one stupendously obvious postulate that accounts for all examples and all results presented in this chapter:

- The human consciousness (or the human soul) "survives" death![30]

Related postulates are the following:

- Consciousness can exist independently from (what we call) matter.[31]
- Consciousness can create (human) shapes.
- Pure (matter-free) consciousness can be perceived.
- Ghosts/spirits/hallucinations are real.

Paying attention for a moment to the last postulate — this is what William Blake emphatically stated. In that respect he has rather interesting company. First, we have the scientist and the great mystic, Emanuel Swedenborg (1688–1772), who echoed Blake in his *Arcana Cælestia*: "That I might know that man lives after death, it has been granted me to speak and converse with several persons with whom I had been acquainted during their life in the body, and not merely for a day or week, but for months, and in some instances for nearly a year, as I had been used to do here on earth."[32] And then there is Kurt Gödel (1906–1978), one of the smartest men of the twentieth century, the person who single-handedly *mathematically* proved that there does not exist (and never will exist) the book of all truths. In 1977, a year before he died, he said the following:

"There are spirits which have no body but can communicate and influence the world. They keep [themselves] in the background today and are not known."[33]

Gödel, like Blake and Swedenborg, spoke from his own personal experience. With or without having a personal experience of such kind, it is reasonable to accept that the above five postulates are rational extractions of the many anecdotal cases and laboratory results supporting them. Their only alternatives are Donkin's law (that each and every such case or experimental results is a hoax), and similar mental acrobatics. On the other hand, these five axioms are a monumental shift away from the materialistic paradigm, and accepting a single one of them means reaching the point of no return. We are conditioned to resist this, and a personal experience would surely help in overcoming this obstacle. We will work on that too.

We are heading in the direction of *extraordinary*, and we should be prepared for it. In order to illustrate this, we reflect on one of the above examples. Recall Mr G seeing the nude "astral body" of his dying wife floating above her physical body, which at times would shrink in size until it was no larger than eighteen inches. If the forms created by our pure consciousness are not constrained by matter, which is what this example and our postulates imply, there is no reason why they shouldn't vary in size. So, within this new paradigm, an 18-inch replica of the dying Mrs G is not a miracle. However, in that case neither should a 16-foot replica of Jesus be a miracle too (Section 2.5). And suddenly, the extraordinary becomes mundane!

Chapter 4

OBEs, NDEs, Reincarnation, Possession

The psyche, in part at least, is not dependent on this confinement [body, time, space].

Carl Jung

OBE is a fairly standard acronym for "out of body experience." Similarly, NDE is short for "near-death experience." Reincarnation memories, or previous personality memories, happen when someone *hallucinates* other *past* lives. Possession occurs when a person's mind, character and memories are supplanted by another person's, living or dead.

4.1 OBEs and Hallucinations

It is very difficult to squeeze into one single section a coherent and representative account of such a ubiquitous phenomenon as OBE, with millions of cases, thousands of which are available in print, most of them just too fascinating to easily skip. Even when reduced to fantasies, the stories of unconstrained, thought-powered trips into deeper realities are amazing to behold. The selection problem is made yet more difficult if one accepts — as we do — that OB journeys are *real*, and logically not reducible to concocted brain products.

The ubiquity of the OBE phenomenon is easy to justify — studies show that between 15% and 45% of the general adult population have had OBEs. The phenomenon is also cross-cultural. According to a 1978 study[1] by Sheils, only 3 out of 44 societies of the time did not hold a belief in OBEs.

We will narrow down the scope of this section by excluding OBEs associated with NDEs — these will be covered in the next section. Also, we will stop at the more extravagant OBEs, usually resulting in a materialization of some sort; these will be covered in the subsequent chapters.

In many cases the experiences are dismissed as inconsequential or *unreal* by the experiencers themselves. Even in studies seemingly sympathetic to *mystic* interpretations, doors are almost always kept open for false but dominating materialistic models. Thus, for example, in Sheils' study just mentioned, "belief" may have been used as a catchword to demote genuine experience and knowledge into "superstition."

The out-of-body trips are often one-time experiences, triggered by special configurations of the circumstances, from mundane (say, being very tired and resting) to extreme (say, anticipating a fatal collision). The trips could be short ("pop out — pop in" cases), and they could be very long in duration — some of Emanuel Swedenborg's OBEs lasted several days non-stop. The longest OB trips — those that start at the end of our lives and last "forever" — will not be covered in this section.

Among those who have experienced OB trips there is a relatively small number of initiates who could take targeted OB journeys at will. Most of the time the OB trips are not under conscious control. This is true even in cases of experienced travellers who have had OBEs multiple times. "Multiple times" sometimes means "thousands of times." For example, the South African mathematician, Joseph Hilary Michael Whiteman, meticulously recorded more than 2000 of his spontaneous OBEs.[2]

A typical OBE starts with — what in medicine is called — autoscopy: observing from a distance one's own body. It's hardly worth mentioning that from the viewpoint of institutional medicine the phenomenon is a (medical) hallucination of some sort. By now it should be even more obvious that we will not follow that line of selective thinking.

We will now take a quick tour through OB cases, starting with relatively straightforward accounts, moving to more and more complex anecdotes as we go, and stopping just before the outrageous cases (which we will cover in Chapters 6 and 7).

We start with the great neo-platonic philosopher, Plotinus (203–270 AD), who said (as recorded by his disciples): "Now when the soul is without body it is in absolute control of itself and free, and outside the causation of the physical universe; but when it is brought into body it is no longer in all ways in control." He spoke from experience. "Many times it has happened: Lifted out of the body into myself; becoming external to all other things and self-encentered; beholding a marvelous beauty."

The following narrative comes from the book *Astral Projections* by Oliver Fox (1885–1949), who was a writer and a prolific OB psychonaut. During one of his many OB episodes, Fox tried to raise himself up through thinking:

The effect was truly surprising. Instantly the earth fell from my feet — that was how it seemed to me, because of the suddenness and speed of my ascent. I looked down on my home, now no bigger than a matchbox; the streets were now only thick line separating the houses. [...] Soon the earth was hidden by white clouds. Up and up and up. Velocity ever increasing. [...] I got really frightened. [...] I willed to descend. Instantly the process was reversed.[3]

Some OBErs went much further: Ingo Swann visited the constellation Sagittarius; William Goodlett had a few OBEs taking him to other planets. Here is an account of such a trip by Goodlett, describing a view from Mars:

I'm standing in front of a small hill of dark colored rocks. The area is cold. [...] I'm looking at the sun. It's less than half the size as seen from earth, and is fainter so that it can be looked at directly. [...] I am not a human of earthly origin. My body is small, about four feet high, and very thin. My arms are leaner

than an earthling's, as are my fingers. I do not seem to have
on many clothes, just some straps, but I wear a fedora-type hat
with a wide brim, and buskin-type shoes that reach half-way
to my own knees. I have thick, dark olive-green skin and my
eyes seem to be larger than a human's. The earth shines in the
heavens to my right, and Jupiter is a brilliant glow to my left.[4]

There are two features in the last example that must be noted.
First, there happened a change of the I-perspective, from
normal or human to *extraterrestrial*. This is reminiscent of the
experiences revealed under hypnosis by Robert Mack's patients
or Dolores Cannon's subjects. We will mostly stay away from
this topic in this book. Secondly, the perceived reality changed
in space and in time. A relatively modern terminology for such
a phenomenon is *timeslip*. We will devote a long section (see
Chapter 6) to such a class of mega-hallucinations.

Let's digress for a couple of paragraphs and go on a tangent.
Imagine the above report is factual; that is, imagine there existed
an advanced civilization of humanoids on Mars! Imagine an
apocalyptic event of long ago wiped out the Martian world
of the time! Imagine some of these Martian souls now exist
as earthlings (Boriska case, Section 3, this chapter). Be utterly
unscientific and imagine a small population of Martians escaped
on this planet! Imagine that the ancient Cherokees legend[5]
about the *moon-eyed white* people of Tennessee, who lived in
windowless stone houses and came out only at nighttime, is
a true account of Martian interplanetary settlers! Imagine that
one such settlement of the moon-eyed people was visited by
two persons during a most bizarre late twentieth-century
drive-through timeslip (Strieber case, Chapter 6). Imagine *now*
that all of the above happens now, that there is only a huge
multi-dimensional present, containing all pasts, all futures, all
realities, and all fantasies. Just imagine! "Imagination is a faculty
of perception," say wise Sufi philosophers.

Is it really "just" an imagination? Courtney Brown used his talent in remote viewing[6] to access Martians. Here is how he describes them: "They [the Martians] have no hair and have larger eyes than humans. Their skin is light."[7] What an uncanny resemblance to the Cherokee legend. This comparison becomes even more intriguing when we learn where exactly Brown located the Martians he saw in his mind travels — they were in the deep underground below New Mexico!

Back to the main story, we will now recount one more anecdote combining OBE and timeslip. The source is Pierre Jovanović's book *An Inquiry into the Existence of Guardian Angels*.[8] Fred, the narrator, was driving a car along a virtually empty road, and was passing a truck, when he glimpsed another one coming in his direction, running with no headlights.

I saw the crash and the monstrous hip of crushed metal. I left my body and observed the mutilated corpses. Then I saw the announcement of my death to my mother and the repercussions in my family. I saw the funeral preparations, the newspaper article detailing the fatal accident and, above all, I attended my own funeral. I remember particularly examining all the faces of those present at my burial. I watched it all from outside. At that moment, inexplicably, the steering wheel turned to the left and our car went off the road to end its trip in the desert. I saw and heard the truck pass as if nothing had happened. He didn't even brake.

Again, we digress for a moment to note one more case of perceiving one's own funeral. It was apparently a meditation-induced hallucination that turned out to be veridical. The source is William Howitt's *The History of Supernatural (in all ages and nations and in all churches Christian and pagan, demonstrating a universal fait)*, Volume 1, 1863:

One night he [Baron von Eckartshausen] remained [awake] till twelve o'clock meditating on the power of magic, when suddenly he heard a funeral song. He looked out of the window and saw Roman Catholic priests going before a coffin with burning wax candles in their hands, and reciting prayers. [...] Eckartshausen opened the window and asked, "Whom do they carry here?" A voice replied, "Eckartshausen." "Then," said he, "I must prepare." He awoke his wife, told her what had happened, and within one hour he was dead. [9]

Back to OBEs. The above-mentioned mathematician, Whiteman, perceived simultaneously parallel worlds during a few of his 2000+ OB journeys. He wrote:

The bright light and the open ground in the one space could be perceived, at the same time as the body appeared in bed in the other space. Moreover, by the exercise of a free choice and decision, the former space could be more and more confirmed, so that the latter faded away, and once again I appeared to lie on the open ground in a bright light. [10]

Said Don Juan: "There are worlds upon worlds, right here in front of us." [11]

Here is a veridical OBE; the original source is *The Phenomena of Astral Projection* by Sylvan Muldoon and Hereward Carrington. Mr McBride, an Indiana farmer, suddenly found himself floating in the air:

The next thing I knew I was, preposterous as it sounds, floating in the room [...] wide awake. [...] I saw that I was floating upward through the building. [...] The ceiling and upper floor failed to stop me. [Eventually he traveled to his old home, where his father lived.] I stood at the floor of the bed in which I saw Father reclining. "Father," I said to him. [...] He was watching

*me, for his eyes were fixed upon me and there seemed to be a
look of surprise upon his face.*[12]

After returning back to his body McBride noted the time of the
"vision." In a few days he visited his father, who told him that
he had seen him; the father also had written down the time of his
own vision, and it matched. Considering the encounter from the
father's angle, this is a case of veridical hypnagogic hallucination!

The following intriguing case could have been veridical.
Miss Z., who had regular OBEs, slept for four nights under
observation in the researcher Charles Tart's laboratory.[13]

*On the third night Miss Z. had a dramatic OBE. She seemed to
be flying and found herself at her home in Southern California,
with her sister. Her sister got up from the rocking chair where
she had been sitting and the two of them communicated without
speaking. After a while they both walked into the bedroom and
saw the sister's body lying in bed asleep. Almost as soon as she
realized that it was time to go, the OBE was over and Miss Z.
found herself back in the laboratory.*

Unfortunately, the California sister refused attempted contacts
by Tart.

We note that in many cases the perception of the *astral body*
during OBE matches the descriptions of the *astral bodies* released
by the dying (see Section 1, previous chapter). For example,
here is the beginning of one OBE by the intrepid psychonaut,
Dorothy Eddy:

*I felt a queer kind of sensation, as if I were sort of expanding. I
got up and felt most beautiful light. I looked down on the bed,
and there was my body, lying there with my flannel nightgown
on, and appearing most unwholesome and unappetizing. And
here was I, stark naked, but feeling gloriously light.*[14]

My fondness for the bizarre is met by the following application of OBE to exterminating rats:

> I [Szalay] was bothered by rats in my studio. About one in the morning I heard a scrapping, rose from bed with a flashlight, and saw a rat. Returning to bed, I projected in the air [he could do controlled OBEs!]. I literally chased the rat around the room, and in desperation it ran between two sheets of glass by a wall to hide. I returned to my bed intending to check these events but fell asleep. The following morning I found a dead rat pressed between the glass and also found fur on the edges of the glass.[15]

Apparently, rats could see the "astral bodies," or the spirits of the OB travelers. So could dogs and cats. This used to be a common knowledge. Quoting from a book by Noël Taillepied,[16] first published 1588, page 144: "Sometimes a spirit will be seen in the house, which perceiving, the dogs will take refuge between the legs of their masters and will not be denied, for they greatly fear spirits."

That the visual spectrum of some animals includes spirits, ghosts, astral bodies and similar is yet another strong indication of objectiveness and tangibility of the phenomenon of disembodied consciousness. There are a few laboratory experiments strongly supporting this thesis. In 1973, Dr Robert Morris conducted several experiments with the talented OB traveller Blue Harari, Blue's kitten and a snake. The cat and the snake were put in separate cages and they were carefully observed. The cat consistently meowed, distressed in the cage, but stopped doing that when Blue was having an "oobie period!" During the same periods of Blue's OB travel, the snake assumed attacking posture and gnawed on the cage.[17]

Instead of cats and snakes, the researcher, Karlis Osis, employed talented people. Osis experimented with Alex Tanous (who claimed he could project out of body at will) and with the psychic, Christine Whiting, together with some non-psychic people, who served as receivers. He instructed Tanous to try to project in a room, and the receivers would try to see someone in a room. Here is what Osis said:

> We used human observers in the projection area. When a human observer was not especially psychic, he seemed to "see" nothing. When an experienced psychic was in the projection area, she did see the projectionist at the approximate time of the projection.[18]

In 1972, Morris conducted electromagnetic spectrum analysis of two of Harari's OBEs. During the first of the two OBEs, Morris got a "beautiful result": "The [electromagnetic] burst corresponded to Blue's oobie almost exactly. His oobie was two minutes long and so was the burst."[19]

We now pay attention to near-death experience (NDE).

4.2 NDEs and Hallucinations

A near-death experience (NDE) is an OBE at the time when the experiencer is close to dying or clinically dead, yet survived. It has been estimated that around 5% of the population have had an NDE. Our goal here is not to discuss the main properties of NDEs — there are many books on that subject — but rather to establish the compatibility of the OB perceptions during NDE with what we call the real world, thus proving the relative objectiveness of the phenomenon — "relative" to the material realm. Ironically, one of our main theses in this work is that the *material world* is consensual, hence subjective, which makes almost everything, NDEs included, subjective too. In Chapter 8 we will try to clarify this seeming inconsistency. For the time being, for one section only, we mainly focus on cases of materially *objective* OB percepts during NDEs, ruling out simplistic explanations of the "NDEs are hallucinations caused by lack of oxygen in the brain" sort.

A typical NDE starts with autoscopy and is followed by the feeling of a rapid movement through a dark tunnel, arriving at a place of brilliant shadowless light, where the experiencer is greeted by relatives, guides, or other personalities, communicating with them telepathically rather than in spoken words. In many cases the experiencers undergo a holographic and interactive life review. The mere consistency of these features is a strong indication that we are not dealing here with arbitrary subjective fantasies; for example, I don't remember having seen a single case of reporting a travel along the tunnel when returning to the body.

Explaining the NDEs as social constructs falls flat because of many reasons, above all because unvarying main features of NDEs are being reported by very young children. Consider, for example, the following account: when Mark was nine *months* old, he fell very sick. He recovered and was never told of his sickness. When he was 3 his mother recorded the following conversation, that she subsequently sent to the researcher Dr Melvin Morse:

He sat down beside his dad, and he said: "Dad, do you know what?" And his dad said: "What?" "You know I died." "Oh, you did?" And he said, "Yeah." His dad said, "Well, then what happened?" And he said, "It was really, really dark, daddy, and then it was really, really bright. And I ran and ran, and it didn't hurt anymore." And his dad said, "Where were you running, Mark?" And he said, "Oh, daddy, I was running up there [pointing upward]." And he said he didn't hurt anymore, and the man talked to him. And his dad said, "What kind of words did he say?" And Mark said, "He didn't talk like this [pointing to his mouth], he talked like this [pointing to his head]."[20]

Morse interviewed Mark and was told that the "really dark" place was a tunnel through which the nine-months old Mark crawled (!) up to the light! Why would a small baby adhere to such an obscure visual pattern unless this pattern is sourced in the *objective* reality?

Morse wrote a few books related to children's NDEs, where he presents many case studies and describes some interesting results of his research. In one of his studies, he interviewed 121 children in a control group, all of whom had very serious diseases (for example, one girl was completely paralyzed for three months), and a smaller study group of 12 children who survived cardiac arrest and "looked death in face." Of the 121 children in the control group, NONE had anything resembling an NDE. On top of that Morse went outside the controlled group and interviewed a further 37 children who had been given anesthetic agents, narcotics, Valium, Thorazine, Haildol, Dilantin, antidepressants, mood elevators and pain killers, to see if drugs can induce NDE. And again, NONE of them had anything resembling NDE. So, he asks the rhetorical question, "If near-death experiences are hallucinations, why did [these patients] not have any experience remotely resembling NDE?" On the other hand, most of the 12 children (at least 8 of them) in the study group had at least one of the NDE traits ("being out of their physical bodies, traveling up some sort of a tunnel, seeing a light, visiting with people who describe themselves as being dead, seeing a Being of Light, having a life review, and maybe even deciding consciously to return to their bodies").

We now pay attention to one of the inaugurators of modern NDE research, Elisabeth Kübler-Ross (1926–2004). Dr Kübler-Ross was a saintly woman, who devoted her life to helping people, especially disadvantaged children.

We will recount two cases from Kübler-Ross' files. Here is the first one (*On Children and Death*):

> One child who was almost lost during very critical heart
> surgery shared with her father that she was met by a brother
> with whom she felt so comfortable, it was as if they had known
> each other and shared each other's lives. Yet, she had never
> had a brother. Her father was very moved by his daughter's
> account and confessed that she did have a brother, but he had
> died before she was born.[21]

Which brings us to the marvelous cases of veridical NDEs! The
following is also from Kübler-Ross' work:

> In one case a female NDEer found herself moving through a
> tunnel and approaching the realm of light, she saw a friend
> of hers coming back! As they passed, the friend telepathically
> communicated to her that he had died, but was being "sent
> back." The woman, too, was eventually "sent back" and after
> she had recovered she discovered that her friend had suffered
> a cardiac arrest at approximately the same time of her own
> experience.[22]

There are some extravagantly outrageous cases associated with
Kübler-Ross; we leave them for later (Section 8.4).

We digress for a pretty, short story from my personal files.
About a decade ago, I read all her books that I could find in the
local libraries. I started the last one, *The Wheel of Life*, on February
17, 2012, two days before I was scheduled to fly to Scottsdale,
Arizona, a town I had never heard of before. On the very first
page of the book, I encountered the following sentence: "[…] I sit
today in the flower-filled living room in my home in Scottsdale,
Arizona." What a cute coincidence! If it was a coincidence at all.
Whatever it was, I took the synchronicity seriously. I located her
grave — a marble plate where her ashes have been buried — and
read there a passage from her book to pay her respect.

Given the close connection between OBE and NDE it is not surprising at all that many NDE cases also involve veridical sightings of distant or inaccessible objects. We have a case of a patient who suffered cardiac arrest, and who, during the ensuing NDE-OBE, "thought her way" up and was distracted by a tennis shoe on a ledge by a third-floor window on the outside of the hospital building. The nurse (Kimberly Clark, the author of the article[23]) then managed to locate and retrieve the shoe. She writes that she had to press her face to the screen just to see the ledge. Another shoe-detecting NDE was recorded by Kenneth Ring.[24] The shoe was observed on the roof of a Hartford hospital during the NDE of a patient who was being resuscitated; it was subsequently found there.

Here is a veridical case of an NDE observation of a practically inaccessible object.[25] When Peter Ballbusch was 5 years old, he drowned when an onrush of water swept him into a brook. He had an NDE; like Swedenborg, he saw a river of people of all races being met by their relatives and he thought of his mother. He was carried to a ledge where he met her.

She knelt down and reached her hand out to me. But I could not reach her. As she leaned forward, a black cross slipped out of her blouse and hung on a thin silver chain right in front of my eyes. I was almost blinded by the sparkle of seven stars which flashed from the dark background of the cross.

Peter was eventually rescued; it took the fire brigade half an hour's work to resuscitate him. Later the child told his father about his encounter with his mother. The father listened with an indulgent smile, until the moment when the child mentioned the episode with the cross, when he abruptly left the room. Years later Peter was told that his father had bought the same type of cross to give Peter's mother at her birthday; his mother died three days before her birthday. The father slipped the cross into her coffin; nobody knew about that except himself!

How can one be serious and, at the same time, dismiss the accounts of *blind* anaesthetized people observing from OB states, objects and events that they could not have perceived even in waking state? Consider, for example, the case of an elderly, blind lady who had diabetes and suffered a cardiac arrest, recorded by Dr Brian Weiss (in *Messages from the Masters*[26]), who was at the time the chairman of the psychiatry department of that hospital.

According to her later report, she floated out of her body and stood near the window, watching, as the doctors administered medicines through hastily inserted intravenous tubes. She observed, without any pain whatsoever, as they thumped on her chest and pumped air into her lungs. During the resuscitation, a pen fell out of her doctor's pocket and rolled near the same window where her out-of-body spirit was standing and watching. The doctor eventually walked over, picked up the pen, and put it back in his pocket. [...] A few days later, she told her doctor that she had observed the resuscitation team at work during her cardiac arrest. "No," he soothingly reassured her, "You probably were hallucinating because of the anoxia [lack of oxygen in the brain]. This can happen when the heart stops beating." "But I saw your pen roll over to the window," she replied. Then she described the pen and other details of the resuscitation. The doctor was shocked.

Let's forget that she was blind and pay attention only to the fact that she described the details of the resuscitation procedure. Could that have been a combination of partial knowledge and wild guesses? The answer is an emphatic and astronomically probable "no"! The cardiologist, Michael B. Sabom, interviewed 32 people reporting autoscopy during NDEs related to surgery or other medical procedures.[27] Of these, none made any flagrant error in describing the CPR (cardiopulmonary resuscitation) that they have observed during NDE. He had a control group of 25 people whom he asked to describe what they thought had happened during their own CPR. Of these people (none of which had NDE), 20 made cardinal errors, three gave limited correct descriptions (at least one of them had seen the procedure done on his father), and two did not know. Of the subject group of 32 cases, there were 6 of them who gave very detailed and correct descriptions of the procedure.

In light of Sabom's research that we have just described, the following case is a virtual stamp of authenticity of the NDE phenomenon. In 1991, Pam Reynolds underwent a difficult surgical operation in order to remove a tumor in her brain. For that purpose, she underwent induced clinical death — the blood from her brain was completely drained out while her body temperature was lowered to "the record level" of 15.5º. Her EEG was, of course, flat during the operation — there was no possibility for any brain function during the procedure. Yet, she experienced a full-blown ND (and OB) episode during which she *heard* the conversation between the surgeon and the nurses (related to her veins), and she *saw* and was able to recall verifiable details about her surgery.

Said John Van Luyk, who died for a while during a cardiac arrest: "So far as death is concerned, I can recommend it to anybody."

We recapitulate: so far in this, and in the previous chapter, we exhibited research and anecdotal cases, statistically proving with probability $1 - \varepsilon$, where ε is a ludicrously small positive number, that consciousness could exist and function independent from matter, and that the associated perceptions are not medical hallucinations. Our goal is to put forward a model of the Deep Reality, asserting among other things that almost everything is a hallucination of some sort, caused by the incompleteness of the I-individuality and by the associated awareness constraints.

We have opened Pandora's box, and the rules of the game called "existence" change profoundly. In particular, the phenomena described in the next two sections are but minor consequences of what we established earlier.

4.3 Reincarnation or Hallucinating Other Lives

In July 2011, I was in Seville for one week. One humid day, during siesta hours, while strolling aimlessly along the virtually empty back alleys of the city, we (I and my wife) stumbled upon the Pillars of Hercules: two pillars hauled there in medieval times from an ancient Roman temple. It was then and there that I experienced my only déjà vu episode. Nothing spectacular, just an oppressive and peculiarly dense wave of gloom and melancholy, combined with a strong feeling of having lived there ages ago, a local child slaughtered by marauding foreign soldiers.

Déjà vu cases constitute a minor subcategory of the past-lives-recall phenomenon. Typically, during a déjà vu event the experiencer recognizes the geography of a place she/he visits for the first time. There is an abundance of such cases; we will represent them by the following story that is intriguing even if fictitious. The original source is an article in *Az Est*, a Budapest journal, and it was quoted in the book *Power of Karma* by Alexander Cannon and requoted in *Reincarnation – Based on Facts* by Karl E. Muller.

A Budapest lawyer and his bride on their honeymoon were sailing up the Danube. Suddenly in a stretch between Passau and Regensburg, the bride became excited, and claimed to recognize the shore, which she proved by describing the scenery before it came into sight. As she fell ill, the couple left the steamer at the next stop, a small town, where they saw a doctor.

The bride insisted on walking through the town, which she declared she knew, saying there was a castle on the adjoining hill, and led the way to it. There, an old caretaker showed them round, but she proved she knew all about the castle. Finally, she wanted to see "the locked room." The caretaker said there was such a room, but within living memory no one had entered it. He told them, "I am over sixty and I came here at the age of ten, and the key is lost." The bride made the caretaker bring a bunch of old keys from a place she indicated. She chose the one which was rusty and said: "There are two corpses in the room. That is why the door has been kept locked." On opening the door they saw two skeletons, one on a bed, the other on the floor with a stained dagger beside it. The bride exclaimed, "I was murdered in this room!" and fainted.[28]

The report in *Az Est* ends as follows: "That this event actually happened, and in the way narrated, is vouched for by the most trustworthy authorities, and the record is therefore made public without further comment, as the incidents are regarded as absolute facts by all who have been brought in contact with them." Despite this pre-emptive defense, we have to acknowledge that this case has many weaknesses — for example, the fact that nobody entered the locked room for so many years defies the power of human curiosity. Nevertheless, we leave this anecdote as is — we present it as a sacrificial offering to the so-called debunkers fishing for weak spots in the narrative. With its inclusion we also make the following two points. Firstly, our argument does not rest on any single case that we present here. Second, by using our normal, waking experience to judge the genuineness of supernormal events, we are doing nothing more than reinforcing our sense of belonging to the realm of normal experiences, therefore making it more difficult to expand our perceptions.

Before we proceed let's make a note regarding the forthcoming terminology. The phrases "past life recall," "reincarnation," "previous personality," "regression," "progression," all refer to the linearity of time, which I do not accept at all, since it is incompatible with both anecdotal and scientific evidence to the contrary. To distance myself from the established meaning I will continue to occasionally use italics, as in *"past* life."

There are two main classes of people who recall and identify with *past* life personalities: children (spontaneous recalls) and people who have undergone hypnotic regressions (induced recalls). Our first example is of a different type: Augustin Lesage (1876–1965), a "psychic artist" from 1912, painted more than 700 pictures.[29] When he visited Karnak in Egypt, he had strong impressions of knowing it. He was especially overwhelmed with happy emotions when he came close to the grave of the 1500 BC Egyptian painter Mena. On the wall beside the grave there was a painting of Mena strongly resembling one of Lesage's paintings. Lesage told a friend that he would see if he had been Mena in a previous incarnation after he died (!). "This friend, some time after Lesage's death, received a message through a medium: 'Tell A.X. that it is so. I was Mena.' The medium had never met Lesage, nor heard any details of his experience. A.X. was **not present** when the message came." A beautiful little case! Mediums and other clairvoyant people with certain talents, will be featured more prominently in Chapters 7 and 8.

In my mind I have a longish list of "things to do after I die," and one of the items is, you have guessed it, checking on a few *previous* incarnations.

Back to the main story. There are hundreds, if not thousands of well-researched and convincing cases of very young children spontaneously recalling verifiable details of their past live existences. In a society where logic, integrity, and truth are paramount, the results obtained by researching only these types of occurrences rule out simplistic materialistic explanations.

Especially noteworthy is the work of Ian Stevenson, whose meticulous investigations of previous personalities (his terminology) among children cover firsthand cases stretching from India and Sri Lanka, through the Middle East and Europe, to North America. In Europe only he recorded more that 500 cases, a selection of 32 of which is presented in his book *European Cases of the Reincarnation Type*.[30] One example is the case of the girl Nadége Jegou, who recalled spontaneously and starting at an early age idiosyncratic details of the life of her maternal uncle, Lionel, who died in a motorcycle accident a few years before she was born.[31] Stevenson researched this case when Nadége was still a very young child and meticulously noted various details. Here are a few of them:

- When shown a photo of Lionel she claimed it was herself.
- When the TV showed a scene in Paris, she exclaimed that that was the place where "Mama works." The scene showed the neighborhood of the place where her grandmother used to work and Nadége had never seen or been there.
- When Lionel's folding bed was once opened, she said that she "slept there when she was small." When she was told that she never slept there, she replied, "before I was small."
- She once said, "When I was Lionel, I used to buy carembars," and then she asked her grandmother, "What are carembars?" (They are candies that Lionel liked.)
- Nadége often and spontaneously made a grimace by protruding her lower lip; Lionel used to do the same. Many more such little corroborative details are given by Stevenson.

We mention the case[32] of the 6-year-old Turkish boy, Kemal Atasoy, who claimed PP (previous personality) as an Armenian leather dealer who lived in Istanbul. The case was investigated by the Australian psychologist, Jürgen Keil, who went to Istanbul in 1997 and confirmed many of the claims of the boy, including the family name of the dealer (Karakash), the location of the house where he lived (the boy mentioned a well know Turkish woman who lived nearby), that he had three kids with "Greek names" (his wife was Greek) etc. Karakash lived a quiet life, and died in 1940 or 1941, some 50 years before Kemal was born. It took Keil a lot of effort and luck to identify an old man who remembered his existence, then follow the lead and confirm the details with local historians.

The distance between Kemal's native town and Istanbul, where his previous personality Karakash lived, is around 500 miles (800 kilometers). The minimum distance between Boriska Kipriyanovich and the place of existence of his previous personality is about 33.9 million miles (54.6 million kilometers). Boriska, born in 1996, was a Russian boy; his previous personality, born, say, a million years ago, was a Martian.[33]

Boriska was, what is sometimes called, an indigo child; he started talking at four months, and at eighteen months he was already able to read. At two he was recounting his life as a Martian humanoid, an inhabitant of an advanced society where there was no dying from old age. As a Martian he was a pilot of a sophisticated spaceship, and with it he travelled to Earth in an instant. He witnessed the cataclysmic sinking of Lemuria, and he, as Boriska, still felt guilty for not being able to save his Lemurian friends. As he grew older, he started providing technical details. He said that the spacecraft consisted of six layers, including a crucial, chargeable magnetic layer that could be programmed for travel anywhere in the universe. According to Boriska, a nuclear apocalypse caused the loss of the Martian atmosphere and wiped out the Martian civilization; the survivors dispersed to other planets or moved underground.

What to make out of this uncorroborated story? Should we dismiss it as a pure fantasy? Not so fast! So far, we have exhibited a proof, probabilistically speaking, that our inner consciousness (soul, spirit, atma, ka, akh, the self, ...) transcends matter, space and time. The possibility, then, that this inner consciousness could make a trip from the body of a Martian to the body of an Earthling becomes a minor corollary. And, yes, going out on a limb doesn't worry me much — Mars did harbor a very ancient civilization. It is just a question of time before this truth is revealed.

From this point of view, Boriska's PP story becomes prosaic much more than extravagant. For more outlandish episodes of migrating consciousnesses — yet still simple corollaries of the fact that *soul* transcends matter — I recommend the works of the Harvard psychiatry professor, John E. Mack (1929–2004),[34] and then perhaps you will be ready to earnestly test your intellectual freedom from the prescribed reality-paradigm with real eyebrow raisers, as are the books by the hypnotherapist Dolores Cannon (1931–2014).[35]

Some of Cannon's subjects were hypnotically regressed into existences as formless energies; one of them operated a spaceship with no inside or outside — what in mathematics is known as a non-orientable manifold. Since such cases are not verifiable by regular methods, we will stick with *ordinary* and present a few convincing cases of hypnotic regressions into past human lives. We will only deal with veridical instances of the phenomenon; cases that cannot be washed away as hallucinated products of hypnotic suggestibility or as cryptomnesia (the return of a forgotten memory without being recognized as such).

Let us start with a weakly veridical but rather entertaining example. It comes from the book *Messages from the Masters* by Brian Weiss:

On the other occasion, a famous Chinese physician visited me in Miami and demonstrated a powerful Qi Gong healing. In return, she requested a past-life regression, and I agreed. She also spoke no English, but she traveled with a translator. She went under very deeply. Within a few minutes she was vividly recalling a past-life scene in San Francisco more than one hundred years ago. She began speaking fluent English during the recall. The interpreter, a trained professional, did not miss a bit. He immediately turned to me and began translating into Chinese. I stared at him for a moment, and then pointed out that he did not need to do that. The shocked look on his face conveyed his realization. He knew she could not speak a word in English.[36]

The preceding example of *xenoglossy* — the ability to speak unknown languages — *can* be explained away as a case of cryptomnesia; the Chinese physician must have been exposed to some English in her life. But the following example cannot be washed away that easily. The source is Ian Stevenson's 1974 article *Xenoglossy: A Review and Report of a Case.*[37]

[Certain Philadelphia doctor regressed his wife "Lydia" to another time. During the next session there were two doctors to observe the regression.] Suddenly Lydia began to talk — not in sentences, but in words and occasional phrases. Some of it was in broken English, but much of it was in language that nobody present could understand. Her voice, moreover, was deep and masculine. [She — or he — identified herself as a man, Jensen Jacoby] [...]. She told of living in a small village in Sweden some three centuries ago. The session was tape-recorded [...]. Swedish linguists were called to translate "Jensen's" statements. In the later session "he" spoke almost exclusively in medieval Swedish, a language totally foreign to Lydia. [They asked "him" question in Swedish, and "he" replied

in Swedish.][...] Certain objects were brought in while Lydia was entranced. She was asked to open her eyes and identify the objects. As Jensen, she correctly identified, in Swedish, a model of a seventeenth-century Swedish ship, a wooden container used for measuring grain, a bow and arrow, and poppy seeds. She did not, however, know how to use modern tools — for example, pliers.

We note that we are not interested in xenoglossy per se; it matters to us only as a veridical element in support of the genuineness of the underlying past lives phenomena.

Here is one more case; this one *cannot* be dismissed as an instance of cryptomnesia. The psychiatrist and professor, Joel L. Whitton, regressed "Harold" into a past life as Xando in Mesopotamia.[38] Harold/Xando was asked to write some English words in the language of the day. He wrote something in some kind of Arabian script. That was given to an expert in ancient Persian and Iranian languages who maintained that the "squiggles" were an authentic representation of a long extinct language called Sassanid Pahlavi (used in Mesopotamia between AD 226 and 651 and which bears no relation to modern Iranian).

We will devote the rest of this section to the following beautiful case study. The researcher is Linda Tarazi, and her results were published in the article titled "An Unusual Case of Hypnotic Regression with Some Unexplained Content."[39]

In the 1970s, a middle-aged woman called Laurel Dilmen (LD) joined an amateur hypnotists club in order to control her weight and headaches. Some of the members of the club started experimenting with past-life regression. LD was regressed to lives in tribal Africa, Sparta, ancient Egypt, sixteenth-century Spain and seventeenth-century England. All except one of these lives seemed not to affect her present state — the only exception was the sixteenth-century Spanish life as Antonia; for some reason she resisted returning back to that experience. The hypnotist at the time was a Dutch man who asked "Antonia" questions concerning Dutch history — the Netherlands was a Spanish possession at the time. She showed knowledge that went way beyond ordinary. Though LD did not speak Spanish, Antonia pronounced Spanish words and names correctly according to the Spanish speaking people in the group. All this made many of the people of the group believe that her memories as Antonia were genuine.

Soon after, LD drifted away from the group. Three years later she came to Tarazi — who is/was a therapist — asking for personal help; LD was starting to have dreams and daytime flashbacks as Antonia! Eventually LD underwent at least 36 tape-recorded hypnotic sessions with Tarazi, spread over 3 years.

It is worth noting, as Tarazi points out, that during the regressions LD was, in general, not concerned with historical facts; they were only side stories to her personal account as Antonia. The historically verifiable tidbits were only brought out after intensive questioning by the hypnotist, and that seemed to annoy LD. Apparently, LD did not have an axe to grind — the info she gave was a response to requests by the hypnotists, not show-offing material! Quoting the article, "Her [LD's] chief concern was not that her story be believed, but that the love that she had experienced be understood."

The detailed life story of Antonia (Michella Maria Ruiz de Prado) is given in the article; moreover, Tarazi wrote a very interesting, marginally fictionalized book based on Antonia's life.[40] We will just briefly summarize.

Antonia was born in 1555 on the island of Hispaniola, in the part that is now the Dominican Republic. As a young woman she moved to Germany, then England, and eventually, when she was in her late twenties, settling in Cuenca, Spain. She did have close encounters with the Inquisition — not portrayed Hollywood-style as a group of blood-thirsty people — but managed to come out of it safely. At 29 she was still a virgin when forcibly taken by an Inquisitor whom she secretly adored. Eventually his lust changed into a deep love, which was reciprocated by Antonia. She had a son by him. She died around 1588 on her way back from Lima, Peru.

Tarazi classifies some of the historical statements that LD/Antonia said under hypnosis as belonging to a well-read person, as LD "might have been." Then she gives a list of specialized facts that "could be located" in American libraries. Such include, say, the date of the "Edict of Faith on the Island of Hispaniola", or "names of priests executed in England in 1581 and 1582" (at the time Antonia was in England). Such not so easily accessible facts could easily be classified as a good-enough proof of the authenticity of the Antonia phenomenon. But, of course, there was much more!

An account of LD's first regression (performed by the Dutch hypnotist) is interesting in itself. She immediately — as Antonia — gave historical details, mainly pertaining to the Spanish rule of the Low Lands, and usually as answers to questions by the Dutch. For example, she said that Don Fernando de Toledo was the Spanish Governor (of the Netherlands). "The hypnotist told her she was wrong — the Duke of Alva was governor then. She replied: 'Of Course. That is his title. I gave his name.' Tarazi comments that even some history books neglect to give his name. It was the extreme accuracy of the numerous details that affected Antonia's life, together with relative ignorance of contemporary events unrelated to it, that presented such an intriguing contrast from the very first session."

In the second session with Tarazi, LD mentioned that Antonia drowned during her return voyage after visiting her uncle, Juan Ruiz de Prado, in Peru. Her uncle's name, and the name of a certain Inquisitor in Peru, were located years later in a very obscure Spanish book published in 1887.

Amazingly, some of the pronouncements of Antonia contradicting the authorities in Spain later proved to be correct! For example, her description of the building that housed the Tribunal of the Inquisition contradicted the information given by the Government Tourist Office of Cuenca. Eventually it was dug up from an obscure book published in 1944 that Antonia was right, and that the tribunal had moved to another building four months before she arrived in Spain. Another such initial contradiction with official information was Antonia's claim of the existence of a college in Cuenca, proven to be correct by digging deep into a rare seven-volume work in Spanish published in 1925. Finally, Antonia claimed that there were two Inquisitors in Cuenca, while knowledgeable professors in Spain claimed that there were always three Inquisitors in the Tribunal. Tarazi went to Spain in 1989 and checked the Episcopal Archives where she confirmed that Antonia was correct, since there was an exception to the general rule of three Inquisitors in the Tribunal for the brief period of time (1584–1588) she was in Cuenca. It is worth pointing out again that Antonia also gave the correct names of these two Inquisitors!

There is, of course, much more in the article. It is interesting that Tarazi tried to extract from LD, when the latter was under hypnosis, some information regarding the potential mundane source of her accounts as Antonia. Antonia, or LD under hypnosis, denied it!

The thoroughness of Tarazi's research is remarkable: for example, she gathered data about everyone she could trace — be it playmates, neighbors, or any other people she knew — having any links to LD during her 45 years of life in 16 different places!

We have presented but a small sample of the huge library of reincarnation cases, all of them together establishing with virtual certainty the genuineness of the underlying phenomenon. The associated medical research also supports this thesis. For example, it has been shown that past-life regressions affect not only memories of the subjects, but their bodies as well: brain wave patterns, blood pressure and heart rate change from one previous life to another.[41] One can then say that during hypnotic regressions the subjects do not merely relive memories of other personalities, rather they become other personalities. Which brings us to the following section.

4.4 Possession: Hallucinations of Becoming Someone Else

Ina and Basil are good friends of mine. They both have had numerous OBEs. Basil dismisses them as hallucinations of no importance. Ina takes them much more seriously. In fact, she is outright scared of them. At times, after a tiresome day, she would have a spontaneous OBE, seeing her limp body sitting on a sofa. She would, then, immediately recoil back to her body; she was afraid that someone else would occupy her soul-less body while she was out. That is, she was afraid of *possession*.

Was she being irrational? Not really. If a consciousness can leave a body, then the proposition that another consciousness could step in and ensoul or re-animate it, is not farfetched. Existence of consciousness without material base — which is what we have been asserting all along — normalizes the possession phenomenon. The accounts presented in this section are mostly given as corollaries, not as a veridical support of our theory.

Possession — supplanting one character or ego-consciousness with another in the same body — can happen in many ways. It can be voluntary or not; it can be benevolent or malevolent; it can be momentary, or it could last a lifetime ... we will exhibit examples of each.

It is hardly worth mentioning that the dismissive medical orthodoxy reduces the whole gamut of possession manifestations to either epileptic fits (that include hallucinations) or to multiple personality syndrome.

A brief but very acute description of what exactly happens during an invited possession is given in *Arcana Coelestia*,[42] (page 6212), by the amazing psychonaut, Emanuel Swedenborg (1688–1772).[43]

> *It is known from the Word [Bible] that there was an influx from the world of spirits and from heaven into the prophets [...]. And as I desired to know in what manner these men were actuated by spirits, I was shown by means of a living experience. To this end I was for a whole night possessed by spirits, who took such possession of my body that I had only a very obscure sensation that it was my own body.*

Another account of a similar sort is given by the writer, editor and skeptic Alan Vaughan, who, while experimenting with a Ouija board, made contact with the spirit of someone called Nada.[44]

> *Marveling at the strong force she exerted on the planchette, I telephoned [...] a friend's house to ask him to come over and observe this amazing phenomenon. There was no answer. And then I did the stupidest thing in my life. I asked Nada to come into my body and guide me to where the friend was.*
> *No sooner were the words out of my mouth than I felt a strange sensation in my brain. [...] Guided by Nada I left the house in search of the friend. My body became a puppet.*

Channelers are masters of invited voluntary possessions. Channelers are defined as conduits to "heavenly," nonmaterial intelligences, human or not. Typically, a channeler would go into a trance, its consciousness would leave the body, and another consciousness would take control. In many cases these visitors from the nonmaterial realms convey exquisite wisdom; however, that is another topic.

Sticking with the subject of this section, we will give another example. It is a combination of possession and levitation; the latter will be more thoroughly covered in Chapters 7 and 8. In 1928, Father Reisinger performed an exorcism of the demons invading the body of a possessed lady; the ordeal lasted 20 days. We read:

Before the nuns who surrounded Anna [the possessed lady] could intervene, the trembling woman on the bed was suddenly carried through the air to the doorway of the high-ceiling room. As if she had been filled with helium, the bedeviled woman remained fixed in midair.[45]

More cases of levitation will be given in Chapters 7 and 8.

The reader should not assume that exorcism is an obscure ritual, practiced mostly in bygone times. In 2018, Pope Francis convened an exorcism workshop at the Vatican that was attended by 250 priests from 51 countries. This was done as a response to the mysterious increase of requests for exorcism all over the world, including the USA.

Uninvited possessions can sometimes produce rather bizarre twists. In his book *30 Years Among the Dead*, published 1924, the physician Carl A. Wickland recounts the following remarkable incident that happened in 1894; at that time he was a student of medicine. One day, after dissecting a leg of the body of a man of about 60, he returned home.[46]

[I] had scarcely entered the door when my wife [who did not know about his dissection] was apparently taken with a sudden illness and complaining of feeling strange, staggered as though

about to fall. As I placed my hand on her shoulder, she drew herself up and became entranced by a foreign intelligence who said, with threatening gestures:

"What do you mean by cutting me?"

I answered that I was not aware of cutting anyone, but the spirit angrily replied:

"Of course you are! You are cutting on my leg."

Realizing that the spirit owner of the body on which I had been operating had followed me home I began to parley with him, first placing my wife in a chair.

To this the spirit vigorously objected, saying that I had no business to touch him. To my answer that I had a right to touch my own wife the entity replied:

"Your wife! What are you talking about! I am no woman — I am a man."

I explained that he has passed out of his physical body and was controlling the body of my wife and that his spirit was here and his body at the college. When he finally seemed to realize this, I said:

"Suppose I were now cutting on your body at the college — that could not kill you, since you yourself are here."

The spirit admitted that this seemed reasonable and said:

"I guess I must be what they call 'dead,' so I won't have any more use of my old body. If you can learn anything by cutting on it, go ahead and cut away."

Then he added suddenly: "Say, Mister, give me a chew of tobacco." I told him that I had none, and then he begged for a pipe, saying:

"I'm dying for a smoke."

"I am dying for a smoke," said the dead man!

The case of Lurancy Vennum was first published in the pamphlet titled *Watseka Wander*, 1878.[47] The author was Lurancy's physician, Dr E.W. Stevens, who first saw her on 1 February

1878. About six and a half months earlier, Lurancy had a fit that lasted five hours, after which she told the family that she could see and communicate with spirits. At the time Dr Stevens came (as an alternative to sending her to a lunatic asylum), she was "possessed by demons." Dr Stevens managed to hypnotize her, which resulted in "free communication with the sane and happy mind of Lurancy Vennum herself, who conversed with the grace and sweetness of an angel, declaring herself to be in heaven." Lurancy from heaven told Stevens that an angel wanted to come, and that her name was Mary Roff. Several hours later Mary came into Lurancy's body, immediately requesting to go home. Mary Roff's family lived on the opposite end of the city of Watseka. The original Mary died as a young girl when Lurancy was 1 year old. Mary, now in control of Lurancy's body, went to live with thee Roff family for three months and seven days, providing ample evidence of her knowledge of idiosyncratic details of the life of her spiritual prototype. On May 7 1878, Mary in tears told her stepmother that Lurancy Vennum was coming back. "She sat down, closed her eyes and in a few moments the change took place, and [completely cured] Lurancy had control of her own body." A beautiful case indeed!

The following lovely *story* comes from the vast library of ancient Indian records. It is about the eighth-century yogi, Sarikara, who was once challenged to a debate by Ubhayabhara, a woman who was an expert in *kamasastra* (kamasutra), the system of knowledge of human sexuality. Now, Sarikara wanted to prepare himself well for the debate, and a problem arose. He needed practical knowledge and he did not want to compromise his celibacy:

After learning, then double-checking, that a certain local king named Amaraka had died, he went to the edge of the town and left his body, devoid of consciousness, in the hollow of a boulder inside a cave. He saw to it that his inert body was guarded by

*his disciples, and then used his yogic powers to enter the body
of the dead king through the [...] aperture in the crown of the
head, with his subtle body.*[48]

Sarikara took over the king's body and started working on his
preparation for the debate: "[...] he joined with his wife, her
face with his, her chest with his, her navel with his, each of
her parts joined with those of his, finding her secret places."
Sarikara, now in the body of the king, seemed to have liked
his preparation so much that he forgot about his original body
waiting in the cave. That made his disciples apprehensive. At
the same time, the ministers of the kingdom suspected that the
dead king had been possessed and gave orders that all lifeless
bodies must be cremated. At the climax of the story, Sarikara's
disciples managed to gain access to him (as a king), and induced
him through music to make him regain his senses and return to
his body, just at the time it was being set to be cremated.

There are a few cases on record of ultimate, lifelong
possessions. We will devote some attention to the following
well-researched example. Our source is the article "The Case
of Iris Ferczády – a Stolen Life," by Mary Rose Barrington,
Peter Mulacz and Tittus Rivas, *Journal of the Society of Psychical
Research*, April 2005. It starts as follows: "In 1933 a 16-year-old
well-educated Hungarian girl, Iris Ferczády, who had dabbled
extensively in mediumship, suddenly underwent a drastic
personality change, claiming to be re-born Lucia, a 40-year-
old Spanish working woman said by her to have died earlier
that year."

Early researchers (during the 1930s) concluded that Iris had
not learned Spanish by any normal way (for she had not even
associated with Spanish speaking people), but some postulated
that she had acquired "a convincing grasp of the [Spanish]
language by telepathy." It is interesting to observe the calculated

approach of skeptics: if a miracle cannot be explained, then the next best thing is reducing it to a *smaller* miracle, appearing susceptible to future materialistic explanations!

Iris was interviewed by the Hungarian ambassador in Spain [...]. He was unimpressed by her mastery of Spanish and her knowledge of Madrid. [...] Ido Ruttkay who has been the Hungarian Consul General in Barcelona, showed her 10 engravings of Madrid; he said that she was "wrong about everything and had never been in Madrid." Very interesting indeed, when viewed from a certain angle — perhaps her past was disjointed from Ruttkay's [we will come back to this point in Chapter 6]. But Ruttkay was surprised by Lucia, since she showed "familiarity with things that he did not expect from a non-Spaniard. She was shown a contraption consisting of a metal pot suspended within a tripod, and she approached it saying, 'Bracero,' and placed her hands over it. The appliance was in fact a brasier [or brazier, a portable heater consisting of a pan or stand for holding lighted coals], and when the pot was filled with hot coal it was used as a heating device by impoverished Spaniards. [...] A Hungarian girl who was not familiar with these articles would probably have assumed the Brasero to be used for cooking rather than hand-warming. [...]" It was abundantly obvious that Lucia was authentic.

That Lucia's past was out of the thin past-present-future line of the linear time (more about this in Chapter 6), is also suggested by the following episode:

At a soirée attended by [a certain Hungarian businessman who spent 15 years in Spain, and] 40 other guests, Lucia was persuaded to speak after dinner, and asked to describe a bullfight. She did this with great enthusiasm but surprised everyone by describing the toreador provoking the bull with a lilac-coloured cape, and dispatching it with a dagger. Her listeners all shouted the cape would be red and the weapon would be a sword. Unabashed, Lucia insisted that she was correctly reporting what she had seen. It is a very curious episode, because anyone who had read up on the subject, and indeed almost anyone who had ever heard of bullfighting, would know about the red cape.

The authors of the article unsuccessfully tried to trace Lucia's family in Spain (which does not imply anything), and they interviewed Iris-Lucia in 1998 (she was 81 at the time):

We asked her how she felt about displacing Iris from her own life. Lucia found it difficult to deal with this and shed some tears. She pointed out that she had not asked to supplant Iris, and had not in fact asked to be re-born at all. She remembered floating happily in space, rather like a small boat in water, in a state of contentment, and then suddenly there she was in the body of this attractive young girl, a virgin again, as she put it, after having given birth to 14 children; and when she looked down she saw lovely young hands, not the worn-out hands of a 43-year-old washer woman. She was pleased with her new body, and she felt that the ousting of Iris was not her fault.

Very interesting!

Lucia mentions the condition of her consciousness before entering Iris' body ("floating happily in space, rather like a small boat in water, in a state of contentment"). This is the *bardo* state, the state of the I-consciousness *between* two consecutive *lives*, which we have tangentially encountered many times in the last two chapters. It fits well in the overall scheme of the deeper reality that we are trying to postulate. We will continue to keep an eye on this subject.

4.5 Reality, Step 4: Facing the Consequences

There exists in mathematics an innocent looking postulate called the Axiom of Choice. In the popular, watered-down version it can be stated as follows: given a (possibly infinite) collection of baskets with apples, we can, theoretically speaking, choose exactly one apple from each of the baskets. Almost the entire (pure) mathematics is based on this "intuitively obvious" and innocuously looking statement. Yet, it implies some extremely counterintuitive theorems, as, for example, the Banach-Tarski paradox: a sphere *can* be cut into six pieces that could be reassembled, without changing the mutual distances between the points, into two spheres of the same radius as the original sphere. There is no way around this conundrum: if one accepts an axiom or a claim, in this case the Axiom of Choice, then all of its consequences, however exotic or absurd they may appear to be, must be accepted too.

In the previous two chapters we have exhibited a probabilistic proof of five postulates that we have stated at the end of Chapter 3. The current chapter consists of particular consequences of these five claims. At the same time, we provided further support of the postulates.

Specifically, the claim that consciousness can exist independently from matter, theoretically validates and normalizes OBE and NDE accounts, as well as the phenomena of reincarnation and possession. OBE and NDE events, for example, become logical consequences of that single postulate. Not that such a validation is necessary: the authenticity and the objective reality of NDEs and OBEs, is also supported by the consistency in the narratives, as well as by scientific research. Some examples of the latter were mentioned above; here is one more.

In 1965, Charles Tart investigated Miss Z., a frequent OB traveller. He monitored her during four consecutive nights and got some very interesting results at the times of her OB trips;

very low alpha brain wave activity and, unlike dreaming state, non-existent REM (rapid eye movement). The outcome of one of Tart's experiments with this talented woman is particularly relevant, as it is a clear-cut probabilistic proof of genuineness of OB perception: the woman correctly reported a five-digit number that was on a card lying on a high shelf well out of her visual range.[49]

The number was 25232.

A similar experiment was done more than 2.5 millennia ago, in 550 bc, by Croesus, the king of Lydia, and a detailed account of this event was given by Herodotus. Croesus dispatched messengers to the famous oracle of Delphi. The messengers were to keep track of the days, and on the 100th day following their departure they were to consult the oracle and inquire from her what Croesus was doing at that moment. The oracle of Delphi, a priestess, in a trance induced by consuming some leaves, uttered the words (reduced to a verse by the present holy men) to the effect that King Croesus was cooking lamb and tortoise meat in a brass covered pot. The messengers returned to deliver her response, and Croesus was so impressed with the result that he ordered a "magnificent sacrifice" of 3000 of each kind of sacrificial beast [what a waste of "beasts"] and sent to Delphi a huge golden pot covered with a golden figure of a lion, all of it weighing more than several hundred kilograms.

Scientists call the Delphi oracle's OB feat a case of distant viewing or traveling clairvoyance. Surprisingly, there is a substantial body of scientific research establishing that some people, or that people in general, have the ability to access perception-wise distant points in this space. So much so, that one of the more prominent researchers in the subject of distant viewing, the theoretical and experimental physicist Harold Puthoff, who was the first director of a distant viewing project funded by the CIA (1970s until 1980s), stated flatly that there is "considerable scientific evidence for the reality of remote viewing as a genuine human capacity." Puthoff said that this claim is backed by "266 highly detailed technical papers and reports I worked on, some of which remain classified [30 years after their production]" and that "there was so much good data and it was so damn compelling."[50]

Puthoff then described to the interviewer (the psychoanalyst, Elizabeth Lloyd Mayer) an experiment that was "mind-blowing." The CIA themselves chose the targets that no one in the scientific team knew about. The distant viewers, Ingo Swan and the police officer, Pat Price, the former mentioned earlier, were given just geographic coordinates, and they then described what they *saw*. Pat Price described some cabins and said that "over a ridge there's this really interesting place." Then he described, in great detail, a military site that he named as highly sensitive and surrounded by the heaviest security. At the end it transpired that the CIA gave the location of a staff member's vacation cabin in Virginia, and even the officers giving the data did not know of the existence near the cabin of a highly classified underground government installation. Price even correctly guessed the actual code name of the site: Haystack![51] Puthoff ends his expose with, "That's what I mean by mind-blowing."

We mentioned earlier that Ingo Swan, or rather his consciousness, travelled to a distant star during one of his OBEs. Is his claim too outrageous? Here is what another prominent researcher, the physicist Russell Targ, said on that subject: "We know, for example, that through remote viewing, many individuals can sit quietly, with their eyes closed, and use their psychic abilities to describe accurately activities, events, and geographic locations all over the planet, and **possibly beyond**" [emphasis mine].[52]

Let's stick with that subject for a moment. Ingo Swan published a short book, a hard-to-get samizdat,[53] in which he describes his experience with the CIA. Once, he was given coordinates of a point on the moon; he first perceived darkness and correctly concluded that the point was on the dark side of the moon. Then he saw tracks on the ground. Worried about the $1000 per day that he was promised, and knowing what he saw was "not possible," he asked for a break. After the break he was given other coordinates of a point on the surface of the moon. To his own dismay and to "Axel's" (his CIA handler) amazement, Ingo *saw* huge artificial structures at the place he was remotely viewing.

So then, are there, or were there artificial structures on the dark side of the moon? Of course there are! The whole surface of the moon has been thoroughly mapped by various lunar orbiters, and a sufficient quantity of these photographs can be found over the web. Despite extensive photoshopping (smudging or planting red herrings here and there), there is ample indication supporting the claim that such structures exist. Taken in isolation, this claim may be weak. Not so in the context of thousands of anomalous observations[54] of objects on or around the moon.

The case of the existence of remnants of artificial structures on the surface of Mars is even stronger. Elaborating would take us too much afar; we will just say that the evidence can be found in books and over the web.

The last two paragraphs are only given as a weak support of Boriska's claim that there existed an ancient civilization on Mars. This topic is not important to us here; what we are interested in is his claim that in a previous incarnation he was a Martian. Within the scope of what we have shown in the first four chapters, there is nothing extraordinary in this claim. Since consciousness is free of matter in its native state, it can merge with, or *ensoul* whatever, wherever and whenever.

Within the spectrum of possibilities, Boriska's case of reincarnation is rather trivial. Really interesting cases start with humans exclaiming under hypnosis how happy they were while existing as "solid blocks of blue light."[55] Outrageous cases are within our reach. A wide-opened mind is required.

Here are our old postulates, as stated in Chapter 3:

- The human consciousness survives death.
- Consciousness can exist independently from (what we call) matter.
- Consciousness can create (human) shapes.
- Pure (matter-free) consciousness can be perceived.
- Ghosts/spirits/hallucinations are real.

The material presented in this chapter leads us to the following additional statements:

- If the circumstances are favorable, the pure, disembodied consciousnesses can communicate with embodied consciousnesses.
- A single human consciousness, to the extent it self-perceives as being single, can manifest in different bodies, sometimes (but not always) separated by (what we call) time.

Chapter 5

Group Hallucinations

Remember that in one way or another, your physical existence is the result of mass hallucinations.

Seth, through Jane Roberts

Group hallucinations are characterized by the fact that they are perceived, or experienced in some way, by at least two people simultaneously.

5.1 Carlos and Fatima

It was early morning when the motel shuttle took me to the Madrid-Barajas airport, and there were very few people in the glass-walled waiting room. People watching is my pastime, but this time, the scarcity of targets rendered that too indiscrete, and I decided to bide my time by playing a game of chess with my computer. I did not notice when someone took a seat nearby. I spotted him only after he casually offered to play a game of chess with me. He was a singularly nondescript man, neither short nor tall, neither young nor old, neither handsome nor ugly. I prejudged him to be a patzer, a chess players' slang for a weak player, and refused his offer.

People watchers in general avoid shallow chitchat. However, my anti-chat defenses are useless when the subject is chess, and anyone can penetrate them with that weapon, patzers included. And so, the man and I started a conversation.

He said his name was Carlos, and that he was on his way back to New York. He was an ethnic Portuguese who came to visit his father and to do some business in Catalonia. We chatted about the Catalan language; he mentioned that, due to Moorish influence, Catalan was linguistically closer to Portuguese than to Spanish. I have always been curious about languages and their interconnectedness — the whole planetary linguistic scheme looks like a fascinating organism with eons-old veins running in many directions through various linguistic enclaves. We chatted about Portugal and its history, and we mentioned the apparitions of Fatima. And then it took off.

At that time, I knew the bare story of the "Fatima miracle." In 1917, three poor young shepherds, two girls and a boy, had several encounters with an apparition of the Virgin Mary. The last time that happened, there gathered many people who witnessed some unusual events. Carlos claimed emphatically that there was no difference between the Fatima miracle and the phenomenon of "alien entities." "It is the same stuff," he said flatly, with a tone betraying pained boredom of someone who states the obvious. Carlos then said that the theme of aliens was an everyday occurrence to him. He said that he owned an art gallery, and that he was constantly inundated with requests by people who had had close encounters with alien entities requesting that he displayed their artwork, which consisted mainly of dolls with oversized heads and huge black eyes. He shrugged his shoulders apologetically; he could not help these poor people lest he sacrificed his business.

It was time to board the plane. We decided to stay in touch and Carlos wrote his email address on a piece of paper. The next day, back home, after the bizarreness of the encounter sank in and became apparent, I looked for his email address. I could not find it. It had vanished! The alternative explanation is that I have lost it. Now, that would have been a miracle, since I never lose objects of value.

Naturally, I decided to take a careful look at whatever sources I could find about the Fatima apparitions. There is a good book titled *Heavenly Lights*, written by two Portuguese historians,[1] containing many details and firsthand accounts. One of their sources is the book *A Miraculosa Nuvem de Fumo* (*The Miraculous*

Cloud of Smoke) by Dr Gonçalo Xavier de Almeida Garrett. Dr Garrett was present during the last and the most spectacular Fatima episode. He was, we note in passing, a distinguished mathematician of Portugal during his career, as well as the dean of the University of Coimbra.

Here is a short synopsis of what happened in Fatima in the years 1916–1917. One day in 1916, Jucinta Marto, 6 years old, her brother Francisco Marto, 8 years old, and their cousin Lúcia Santos, 9 years old, shepherded their parents' flocks of sheep in the field called Cova da Iria near Fatima, when they saw an angel of supernatural beauty, who instructed them that they should regularly repeat a certain prayer, then promptly disappeared. In May of the following year the three children saw, descending from the sky in flashes of light, an apparition of a "small, pretty lady" (Lúcia's description). The small, pretty lady was subsequently identified by Jucinta as the Virgin Mary. The encounter was repeated several times at pre-appointed days, in the presence of an exponentially increasing number of eyewitnesses.

Nobody except the two girls actually saw — or hallucinated — the apparition of the pretty lady, or the Virgin Mary, during these repeated encounters; Francisco could only hear the Virgin Mary's voice, while the other eyewitnesses perceived various other wonders in the sky and on the ground. For example, we read of people perceiving loud thunders in fair skies, subterranean humming and hissing noises, rains of flowers, and bizarre misty clouds, sometimes changing colors, and moving seemingly purposefully in different directions. Others saw the sun losing intensity. A certain Maria Carreira reported that "the trees seemed not to have branches or leaves, but only flowers. The ground was divided into squares, each one in a different color."[2] Many people saw a luminous globe gliding through the sky. For example (eyewitness account, August 1917[3]): "...

a silver globe appeared making small circuits and appearing to cut, here and there, through the clouds."

The grand event happened on October 13, 1917, during the last appointment of the three children with the Virgin Mary. The crowd in the vicinity of this event was estimated to have consisted of at least 50,000 people. The main show, the coming of the Virgin Mary, was again only for the three small children. However, the sideshows were not less impressive. A sound of buzzing bees was followed by erratic movements of the clouds, and flashes of colored light ricocheted from one cloud to another. Then the clouds suddenly vanished, and the center stage was taken by a sun, appearing in a bright sky. Some people perceived the sun as a bright disk, "with a distinct and lively border, luminous and lucent."[4] Others saw it changing colors. The mathematician, Dr Garrett, noticed that "the edges of the 'Sun' executed a rotating movement."[5] Then the "sun" descended in a slow, zigzag rhythm of a falling leaf.

This miracle of the Dancing Sun was not confined to the Cova da Iria field. Seven miles away, people were tending to their everyday chores, when they saw it. Here is a quote from a letter by Father Ignace Pereira, written in 1931; that day in 1917 he was attending a class at his school when he and the other students heard some commotion outside and went to investigate:

> Outside the people were gathered in the square, crying, shouting, pointing to the sun, without even hearing the questions poured upon them by our terrified teacher. I stared fixedly at the sun. It seemed pale, without its usual flashing light. It looked like a ball of snow turning upon itself. Then, all of a sudden it seemed to come down in zigzags, threatening to fall upon the earth. Scared out of my wits, I rushed into the crowd. Everyone was crying, expecting the end of the world.

Close to us was a freethinker, a man who had spent all morning laughing at those leaving for Fatima. Now he stood there as if paralyzed, stupefied, his eyes fixed in the sun. I saw him afterwards trembling from head to foot. Finally, raising his eyes to heaven, he fell on his knees in the mud of the road, repeating over and over: "Holy Virgin! Holy Virgin!"[6]

The Fatima event is merely one of many similar supernatural phenomena intermixing religious connotations with other themes and witnessed en masse by multitudes of people. We will only briefly mention a few more such cases. The so-called Welsh Revival manifestations occurred in 1904–1905. This episode started with inner voices experienced by a local preacher, and it evolved into mass sightings of stars and fiery balls that changed their form and moved in different directions, as well as in perceiving heavenly music originating in the sky. In another case, four girls of the village of Garabandal, Spain, had an experience that parallels the experience of the Fatima children: they also met an angel who announced the vision of the Virgin Mary. Their ecstatic visions lasted between 1961 and 1965. In Egypt, in the year 1967, a huge crowd, consisting of estimated 250,000 people during one single day, observed a luminous apparition of a woman, Virgin Mary or not, moving smoothly on the roof of a Coptic church in the suburb of Zeitoun in Cairo. One day the apparition was visible for eight consecutive hours. Photographs and videos of the event, "fake" or "genuine," can easily be found on the internet.

The most frequent method that the deceivers use to sweep the Fatima miracle (and similar events) under the carpet is to amplify the obvious, and to portray the event in strictly religious terms. It is then left to those who could think, and thus pose some threat to the grand materialistic party line, to deduce by "themselves" that the event was no more than a case of mass delusion of frenzied religious zealots. However, it is not so simple. Among the people in the crowd there were many hundreds of secular intellectuals or curiosity driven onlookers. They all, irrespective of their expectations or hopes, perceived a profusion of extraordinary events. As we have noted above, this includes the cases of unsuspecting people, far from the main event, some of whom ridiculed those who left for Fatima, also witnessing supernatural events.

Ironically, interpreting these events as mass illusion, as a superimposed reality, or as induced experiences, may not be far off the mark. Though it has all of the features of a staged spectacle directed by technologically advanced entities, the event might have primarily showcased "technology" of superimposing hallucinatory scenes on top of the *normal* reality. The events described in the rest of this chapter, especially the cases of mass hallucinations, will strongly imply that this interpretation of the Fatima miracle may be the right one.

5.2 Group Hallucinations Induced by Psychedelic Drugs

Psychedelic drugs used to be called hallucinogens, precisely because the experiences they induce are considered to be hallucinations according to the standard definition. However, the cases we will cover indicate that they are more — much more — than a route to experiencing subjective imaginary realms. This is not merely an opinion based on anecdotal evidence; it is a statement based on significant scientific research.

It is also a belief shared by many ancient civilizations. We mention that out of the 1028 hymns appearing in the ancient Sanskrit text *Rig Veda*, 120 are devoted to the mysterious psychoactive plant Soma. Here is one:

We have drunk Soma and become immortal; we have attained the light, the Gods discovered.
Now what may foeman's malice do to harm us? What, O Immortal, mortal man's deception.

What is particularly fascinating about the effects of psychedelics is the consistency in the types of experiences induced by specific drugs in specific circumstances. Let us consider the case of the die-hard skeptic, the physician Ronald K. Siegel.[7] Siegel researched altered states of consciousness and, to his credit, at times he used himself as a guinea pig. He started with the so-called Lily tank — a completely isolated cylindrical enclosure, filled with warm, salted water, used, through sensory deprivation, for inducing various types of *hallucinations*. John C. Lily himself utilized his invention to get in touch with (hallucinated?) aliens. So, Siegel went into a Lily thank, lied still in the water, in total darkness and silence, and patiently waited for whatever might come about. Sure enough, images of various geometric shapes started parading in front of his wide-open eyes. However, his metaphysical trip turned out to be rather short; it ended with a 3D image of a little Buddha with Mickey Mouse ears, holding a pink balloon that read "I am them." The little Buddha, as it transpired, was Siegel's threshold guardian. It produced a shiny golden needle, poked himself in the navel and exploded, which in turn caused startled Siegel to almost choke on salted water. What an appropriately cartoonish finale! His conclusion was trite enough: it was just a simple pseudo-hallucination (a hallucination that failed to fool him).

Siegel was not fooled by peyote either. He took it in a controlled experiment, together with a group of people, and he saw a "black, gauzy curtain with a large human eye in the center surrounded by a symmetrical arrangement of smaller eyes. The eyes were alive, leering."[8] So far nothing fishy, just a bizarre random fantasy. Or was it? He asked the other participants in the peyote experiment about their experiences, and they all described the same image, down to the number of eyes — thirty, plus/minus one or two! Moreover, every psychonaut — Siegel's apt term — including one Indian and one Japanese person, reported the same Caucasian eyes! How could that have happened? Since any objectiveness of the *hallucinated* scenes is a priori excluded, Siegel set himself to finding other explanations. And he found one easily:

> The [eye] demon [is] an archetypal mental image available only to those in the altered state of consciousness, and now unearthed by my team of psychonauts; perhaps members of other hallucinogenic cultures saw it.[9]

Words, words, words! Do they explain anything of substance in this case? Or are they merely an attempt of a pseudo-Jungian evasion, explaining one *miracle* in terms of another that looks more swallowable.

We end our account of Siegel's saga with his encounter with an old shaman during a peyote séance on the slopes of Sierra Madre Occidental, Mexico. He asked the old shaman: "My friends all see the same hallucination of trienta tsikuri [the multi-eyed god]. Why?" We will devote a new paragraph for the shaman's wise reply:

> There are no hallucinations with peyote. There are only truths.[10]

There are truths with DMT too. DMT stands for Dimethyltryptamine; it is a powerful psychedelic, sometimes called the spirit molecule. The two terms meet in the title of the Rick Strassman's book *DMT: The Spirit Molecule*, which will be the source of our next batch of cases of "group" hallucinations under influence of drugs.

Sixty carefully selected subjects of Strassman's experiment were intravenously infused strict doses (0.03–0.4 milligrams per kilogram) of DMT. For fifty-seven of them — or for their consciousnesses — that meant almost instantaneous journeys into other realities. These fifty-seven then reported their hallucinations as they unfolded.

Drugs affect perception, so the subjective experience of "other realities" by the volunteer psychonauts in Strassman's experiment is nothing unusual. What is extraordinary and relevant is the thematic consistency of the reports. There were no thirty-two-eyed gods this time, but rather many interactive encounters with humans and with sentient alien beings! Strassman did not anticipate this. He writes, "contact by our research subjects with these 'beings' made the most disturbing and unexpected type of DMT session."[11]

The psychonauts interacted with "solid" (what does "solid" mean anyway?), and sometimes utterly bizarre alien beings, including "elves [holding] up placards showing me [Karl, one of the subjects] these incredibly beautiful, complex, swirling geometric scenes."[12] Then, in another case "an insect-like thing [that ...] sucked me [Aaron] out of my head into outer space, a black sky with millions of stars,"[13] and "a single Gumby, three feet tall, attending me [Jeremiah]."[14] In another case, a volunteer called Willow got acquainted with winged gremlins. Perhaps the most bizarre of these encounters was the one by Ben, "There were four or five of them. They were on me fast. As crazy as this sounds, they looked like saguaro cactus. [...] They were flexible, fluid, geometric cacti. Not solid. [...] They probed, they really probed."[15]

Others saw fantastic scenes of a different sort. Lucas was led to a spaceship platform; Gabe was catapulted through a tunnel into the open space, an event similar to that experienced by people who clinically died, but then lived to tell the story; Dmitri found himself in an alien laboratory, where friendly beings waited for him; Rex encountered beings with elongated heads, and insects that satisfied their hunger for emotions by feeding on his heart.

The perception of oneself was also drastically altered in some cases. For example, Jeremiah, who as we mentioned above, interacted with Gumby, perceived himself as a baby-Gumby; and Sara was an energy form. Being a kind of liberated energy form is a fairly frequent self-description of NDE-ers and OBE-ers.

Why did such types of encounters happen so many times during Strassman's experiment? And why is there such an evident similarity between the DMT-induced experiences, and the perceptions of some deeply hypnotized people in other studies, particularly the study of the Harvard professor, John Mack?

And why did almost all of these 57 people report strong "vibrations" brought on by DMT; why has this phenomenon of fast-pulsating vibrations also been reported many times during other types of altered states of consciousness? Partial answers to these questions were given in the previous chapters (recall, for example, our brief journey into String Theory, Section 2.7). We will expand further in Chapter 8.

We finish this section with the following somewhat outlandish example of shamanic usage of psychedelics. It was related by the Evangelical Archbishop of Uppsala, who, at the time (nineteenth century), was sent by his (Swedish) government to investigate "the wide-spread superstitions" in Lapland. In Lapland he was a guest of a well-to-do Laplander, who "had the reputation of being a magician." The Archbishop's mission was supposed to be a secret, but his host apparently knew the purpose of his visit.[16]

In his surprise, the Archbishop made no bones about acknowledging that it was so, and added that this kind of "nonsense" could be reconciled neither with the Christian religion nor with scientific knowledge. To this, Peter Lärdal [the Lap host] replied that he could not prevent His Lordship from doing whatever he considered to be his duty; but that the whole business had nothing to do with "nonsense", and that he was ready to prove it immediately. He would separate his soul from his body, and the Archbishop should tell him where to send it. When he returned he would give proof that he had actually been to that place.

Torn between curiosity and his principles, the Archbishop finally agreed, privately hoping to come on the track of some fraud. He suggested, therefore, that Lärdal's spirit should go to Uppsala and bring back news of his wife.

Meanwhile Lärdal had fetched a pan containing some dried herbs to which he sat fire. He said that he would inhale the flames of the herbs, which would make him lose consciousness. In no circumstances must he be touched, because that would kill him. He would regain consciousness after about an hour.

So, Lärdal lay in his chair for an hour, like a dead man, his face deathly pale. Then he woke up twitching convulsively and described the Archbishop's kitchen in Uppsala. In order to prove that he had been there, he said, he had taken the Lady's wedding ring and had hidden it in the bottom of the coal scuttle; she had taken it off because she was cooking.

The Archbishop wrote to his wife at once and asked her to tell him what she had been doing that morning, and where she had been. The answer arrived a fortnight later. She said she would never forget that morning because she had been making a cake and had therefore taken off her wedding ring and laid it on the table. The ring had disappeared. For a few moments a well-dressed Laplander had been in the kitchen. He had made no reply when she asked his business, and had left without saying anything. This man, she supposed, must have taken the ring.

Later the ring was in fact found at the bottom of the coal scuttle. The shaman's hallucination thus became a reality for a distant witness — and is what allows us to categorize this intriguing case as a group hallucination.

5.3 Group Hallucinations Under Hypnosis

Hypnotically induced group hallucinations are ubiquitous: many of us have seen bazaar conjurors hypnotizing groups of volunteers into flapping and clucking hens. Such cases will not be considered here, though they are not without value since they underline the inherent human potential for changing subjectively perceived realities. We will rather pay attention to attempted scientific experiments, not necessarily conducted under strictly controlled conditions. Such are relatively rare.

First, we note an experiment by French hypnotist, Hector Durville (1849–1923), conducted mostly in his own house. An insert from Scott Rogo's book *Leaving the Body*:

> For one series of tests, he and a colleague hypnotized a female psychic. An observer was placed in another part of the house. The experimenters "sent" their subject's double into the distant room and instructed it to either touch, hit, or pull the observer. These trials were invariably successful. The observer, who had no idea what the subject's double would be requested to do, often felt blows, touches, or pulls by invisible hands. The phantom double was even visible to some of the people whom Durville used as witnesses.[17]

From the orthodox point of view the subject and the witnesses were just imagining, and the observer's part of this experiment was a case of coincidental tactile hallucinations — case closed.

The next experiment, carried out in England in 1969, is very reminiscent to Strassman's DMT experiment. The quote is from Johannes Von Buttlar's *Journey to Infinity*:

> Four people, under the auspices of a hypnotist, agreed that they would attempt to project themselves on to another planet under hypnosis. A fifth man offered himself as an observer. The

hypnotist put himself [!] and the four people into a deep sleep. The experiment lasted for about two hours. As soon as they awoke, they all, without exchanging a word with each other, immediately wrote an account of their experience[s]. They were all in agreement in their description of a voyage through space, of a strange planet with curious, leafless trees whose branches curled round in tight ring, of lakes of oil, a great red sun, fissured red rock, and vast expanses of desert... The timing of the experiment was verified by the observer, as well as the fact that the hypnotist had put himself and the others into [...] trance and had kept silence.[18]

What made the four volunteers perceive or imagine the same scene? If there are subliminal connections among us, what triggers their manifestations? Is the vision of the particular planetary panorama a function of that particular hypnotist? Why was that particular door to other perceptual realms opened? Are there tangible parameters that define our perception and that can be manipulated in controlled ways? These are questions that apply to this and many other cases we will survey.

The next unique case belongs to this category. In an experiment conducted by Charles Tart, two graduate students in psychology, "Anne" and "Bill", hypnotized each other.

First Anne hypnotized Bill, then Bill Anne, then they both deepened each other's [hypnotic] depth (while both being responsive to Tart). They apparently shared a mutual "experience," during which both were taken through some tunnel that was in some way under control of Bill:

Anne [was] intensely curious as to what lay at the end of the tunnel, the end that Bill would not let [her] reach. [...] About a month after [the second] session, Anne was a subject of a group hypnosis test. As she knew what the induction procedure was, she decided to "go" back to the tunnel and explore it as soon as she was hypnotized but before the suggestibility test items were administered. She found herself running along the tunnel, hurrying to reach the end before the test items. At the end of the tunnel she found a cave, blazing with brilliant white light, and occupied by an old man of angelic appearance. [...] He [...] finally told her, very sternly, that he could not answer her question because Bill was not with her.[19]

How "real" was her experience, and their joint experience? Quoting Charles Tart:

I asked the subjects about their perceived bodies during their experience and found that they were curiously disembodied much of the time. They mentioned having heads or faces but no bodies at times, and Anne reported that they walked through each other sometimes. When Bill commanded Anne to give him her hand so he could lead her back, Anne reported that she had to "crawl back into my body, sort of. It was almost as if [we] were moving around with just heads. When Bill said [to] give him my hand, I had to kind of conjure up a hand."[20]

Conjuring up material forms is often reported by people experiencing out-of-body state of consciousness.

One can loosely classify the thematic consistency of the experiences of the subjects of a specific hypnotist as a group hallucination. For example, many subjects of the psychiatrist John Mack[21] relived experiences of non-human entities, without being goaded in that direction. Some other hypnotists, working with the same types of subjects (people who claimed contact with alien intelligent beings) had no such cases. Our influence over seemingly independent domains of reality is a phenomenon observed by many. For example, it has been observed that Freud's subjects often had Freudian dreams, typical Jung patients reported their Jungian dreams. In parapsychology this is called the sheep-goat hypothesis — the results of an experiment are strongly affected by the beliefs of both the subjects and the experimenter. All of this is a corollary of the ancient thesis that *everything* is interconnected. That is one of the directions of our narrative.

5.4 Spontaneous Mass Hallucinations

Mass hallucination is a single hallucination, experienced simultaneously by a large group of people. That in itself makes it harder to explain away as individual fantasies or percepts not sourced in *reality*.

Where do we start? Firstly, and straight to the point, the whole "material reality" is a mass hallucination of a sort. This is most certainly not an original statement. Ancient Vedic texts allow this interpretation, and there are many thinkers who have arrived at the same conclusion. For example, the French historian and critic Hippolyte Taine (1828–1893) taught that exterior perception is a pure hallucination. The physicist and parapsychology researcher, Charles Tart, (born 1937) concurred.[22] "We are living in mass hypnotism", he once said. However, for the time being we will postpone deeper discussion, still stick with the orthodox definition, and focus on giving examples of mass hallucinations relative to the "material reality" (but I cannot omit the quotes!).

Perhaps the most frequent, even ubiquitous, spontaneous mass hallucination consists of perceiving scenes in the sky. In a number of cases there occurred spontaneous mass hallucinations of *mass* scenes. For the purpose of the theses that we are pursuing here, the primary value of these phenomena is their sheer existence, as well as their extravagance. We also note that there is obvious overlapping between spontaneous mass hallucinations and the so-called timeslips (perceiving non-present scenes); we will pay more attention to this link when we cover timeslips in the next chapter. Meanwhile, we provide some interesting examples, ordered mostly by chronology.

In the year 922, an ambassador from Baghdad visited King Almush (or Almish) of the Volga Bulgars, seeking a military alliance against the Khazars. In his subsequent report he wrote:

A cloud was seen not far from me, red like fire [...]. And [...] [there were] seen in it something like men and horses and in the hands of some figures inside [the cloud], similar to men, [there were] bows, arrows, spears and naked swords.[23]

King Almush then told the ambassador that his forefathers were accustomed to seeing such things.

Let's note before we proceed that we are fully aware of the *normal* explanation of events such as this one (meteorological phenomena and coincidental cloud formations). Normal explanations are rooted in normal reality; the absolute objectivity of the latter is what we challenge here.

Holland, 1182:

Four suns and a score of armed men were seen in the sky and bloody rain fell.[24]

Was the bloody rain also hallucinated? Or should we dismiss this as a fantasy of the chronicler?

On October 23, 1642, there occurred a particularly vicious, day-long battle between the army of King Charles I, commanded by the king's nephew, Prince Rupert, and the forces of the third Earl of Essex. Some 30,000 soldiers fought in a bloody, yet inconclusive clash. A few months later the local people first heard the sound of the battle, then saw the battle re–enacted **in the air!** The following quote was written in 1643 "by a historian of that time," in a pamphlet titled *A Great Wonder in Heaven*:

It came to travelers, astonished and fearful, beheld in the air the same incorporeal soldiers that made these clamours [...] The struggle lasted till two or three in the morning [...] amazing and terrifying the poor men on the road who saw all this at the doorstep to the heaven.

King Charles I, upon hearing of the phenomenon, dispatched several of his officers to investigate. The officers interviewed the witnesses. They also, on two occasions, saw the battle themselves and recognized some of the men who had died in the battle. It is very interesting that they also saw the figure of Prince Rupert, who was alive at that time![25]

An event mentioned in the journal *L'Astronomie*, 1888:

About the first of August, 1888, near Warasdin, [Austro-] Hungary [now Varaždin, Serbia] several divisions of infantry, led by a chief, who waved a flaming sword, had been seen in the sky, three consecutive days.[26]

And here is the American equivalent, first published in *Brooklyn Eagle*, January 18, 1892:

A mirage in the sky of Lewiston, Montana — Indians and hunters alternatively charging and retreating. The Indians were in superior numbers, captured the hunters [and] tied [them] to stakes.[27]

We end this type of mass hallucinations with a truly bizarre batch of cases, spanning an interval of more than a couple of millennia in time. They can hardly be labeled "hallucinations" (with the orthodox meaning), and the reason will soon be clear.

The first act in these narratives is the same in all cases: people on the ground see a sailing ship floating in the sky. The first account of "ships in sky" (that I could find) appears in *History*

(Books XXI–XXII) by Livy, and it is rather laconic: "Phantom ships appeared in the sky." This happened in 218 BC. The later accounts of "ships in sky" were more detailed.

The *belief* in the reality of this phenomenon was so entrenched among the eighth- and ninth-century peasantry of the environs of Lyon, France, that the local archbishop, a Spaniard called Agobard, felt compelled to write a pamphlet titled *Against the Multitude's Absurd Belief Concerning Hail and Thunder*, where he denounced this "foolishness."

In the second act of the drama, the sailors of the sky-ship throw an anchor on the ground, as if the ground is the bottom of a sea. We have two different thirteenth-century sources: the first, *Otio Imperialia* or *Recreation for an Emperor*, was written between the years 1210 and 1214 by Gervase of Tilbury; and the second is a Norwegian manuscript from around the year 1250, called *King's Mirror*. Here is an insert from the first:

A strange event in our own times, which is widely known but none the less a cause of wonder, provides proof of the existence of an upper sea overhead [!]. It occurred on a feast day in Britain, while the people were struggling out of their parish church after hearing high mass. [...] To the people's amazement, a ship's anchor was seen caught on a tombstone within the churchyard walls, with its rope stretching up and hanging in the air. [...] Soon, when their efforts proved vain, the sailors sent one of their number down; using the same technique as our sailors here below, he gripped the anchor-rope and climbed down it, swinging one hand over the other. He had already pulled the anchor free, when he was seized by the bystanders. He then expired in the hands of his captors, suffocated by the humidity of our dense air as if he were drowning at sea. The sailors up above wasted an hour, but then, concluding that their companion had drowned, they cut the rope and sailed away, leaving the anchor behind. And so in memory of this

*event it was fittingly decided that that anchor should be used
to make ironwork for the church door, and it is still there for
all to see.*

A drowned hallucination?!

The Norwegian account differs in the location (Ireland
instead of England), minor details (anchor caught in the arch
above the church door, instead of a tombstone), and the final
outcome (the diving sailor was let go). Otherwise, the two
accounts are same. The standard explanation must then start
with the assumption that the first account is fictional, and
that the second is a modified copy of the first. For those who
subscribe to the orthodox materialistic models of "the reality,"
this is the only reasonable interpretation. That type of reasoning
is precisely what we are challenging.

A variation of the incidents of anchoring sky-sails
happened at least twice in the late nineteenth century. Both
episodes happened in America, both in the spring of 1897.
The first case occurred in March, when an anchor fell from
the sky on the farm of Robert Hibbard, not far from Sioux
City, Iowa:

*On the night in question, he [Hibbard] says he was tramping
about his farm in the moonlight... when suddenly a dark body,
lighted on each side, with a row of what looked like incandescent
lamps, loomed up some distance to the south of him at a height
of perhaps a mile from the ground. He watched it intently until
it was directly over his head. At this point the skipper evidently
decided to turn around. In accomplishing this maneuver the
machine sank considerably. Hibbard did not notice a drag rope
with a grapnel attached which dangled from the rear of the car
until suddenly, as the machine rose again from the ground,
it hooked itself firmly in his trousers and shot away again to
the south. Had it risen to any considerable height, the result,*

Hibbard thinks, would have been disastrous. Either his weight was sufficient to keep it near terra firma, however, or the operator did not care to ascend to a higher level.[28]

About a month later we have the following episode (Merkel, Texas), quoting from Houston's *Daily Post* on April 28, 1897:

Some parties returning from church last night noticed a heavy object dragging along with a rope attached. They followed it until in crossing the railroad it caught on a rail. On looking up they saw what they supposed was the airship. It was not near enough to get an idea of the dimensions. A light could be seen protruding from several windows; one bright light in front [was] like the headlight of a locomotive.

After some ten minutes, a man was seen descending the rope; he came near enough to be plainly seen. He wore a light-blue sailor suit, was small in size. He stopped when he discovered parties at the anchor and cut the rope below him and sailed off in a northeast direction. The anchor is now on exhibition at the blacksmith shop of Elliott and Miller and is attracting the attention of hundreds of people.

How incredibly similar to the two thirteenth-century accounts.

What to make out of these falling anchor incidences? Who pulled the strings in the spectacles? And what is the purpose? Is the repetitiveness of these bizarre anecdotes precisely the main point, a kind of emphasis of the tentativeness of what we perceive as material reality? Did someone want to tell us that our world is a joke?

5.5 Induced Mass Hallucinations

Compared to spontaneous mass hallucinations, induced mass hallucinations are distinguished by the fact that they are ostensibly triggered by human actions.

Let's start with the *Indian Rope Trick*; until a century ago, or less, it was a ubiquitous fakir or yogi performance showcased throughout Asia, and witnessed by hundreds of thousands of people, including the paragons of "scientific skepticism," the ruling class of Western European folk, mainly British. Here is a brief description of the trick from the book *Powers That Be*, by Alexander Cannon. We mention in passing that Cannon was perhaps the first hypnotist to regress people to *previous lives*, that is, to lives preceding birth. He witnessed the event around the year 1930, together with a certain Monsieur Robert of the French Consulate of Indochina. In it a yogi lifts a rope in the air holding only one end of it:

> *Yogi then motions the boy [that accompanied him] to take hold of the rope and to climb [the suspended rope]. [...] The boy climbs the rope and the Yogi appears to follow him up the rope with a knife in his teeth. He gets hold of the boy, cuts him in pieces, and appears to drop the different parts of the body to the ground, where they lie quivering in the dust. The magician then descends the rope, puts the pieces of the body together and then sends the youth up at the top of the rope again.*[29]

In many cases the audience consisted of many thousands of people, almost all of them experiencing the same miracle. The narratives of other witnesses, during other "Indian Rope Trick" events, share the above synopsis, at times with minor differences in details. For example, in the old account by the traveler Edward Melton, the trick, performed by Chinese conjurers in Batavia (Jakarta), in the year 1670, with an audience of some 50,000 people, ended with "limbs creeping together again, and [...] forming a whole man."

How could a single person so nonchalantly induce such bizarre and surreal scenes simultaneously into the minds of 50,000 people? And, if this can be done by relatively low-level initiates like fakirs, then what can superior, human or non-human beings do to our perception of what we think is reality? Was this how the miracle of Fatima happened? Assuming the existence of truly superior entities residing in the realms beyond matter, which we certainly do, do these examples open up the possibility that they could induce global realities directly into the minds of those who choose to experience them? Is the genesis of our reality such?

And, if the material realities are so tentative and malleable, are there dissenters who genuinely perceive realities different from the consensual one? As we will soon see, laboratory results in experimental physics imply that our reality and our perception of it are closely intertwined. That alone makes the affirmative answer to the previous question very reasonable.

Let's stick with anecdotal evidence for a while longer — was the perception of the Indian Rope Trick common to all 50,000 people? No, it was not. The traveler and writer, Timothy Wyllie, whom we have mentioned in the previous chapter, viewed the proceeding of an Indian Rope Trick from a palace overlooking the courtyard where the trick took place.[30] He saw the small kid shimmering up the rigidly vertical rope — which is a trick in itself. For the crowd in the yard the boy disappeared; but from his vintage point Wyllie could see that the boy was clinging to the rope, keeping still and quiet. Was Wyllie just barely out of reach of the fakir's hypnotic spell?

We also have material anecdotal evidence. According to the researcher and physician, Andrija Puharić, during one performance of the Indian Rope Trick, two psychologists, perceiving the same miraculous scenario together with hundreds of other witnesses, surreptitiously took a few photographs of the fakir in action: "When they developed their film, they saw to their astonishment the fakir and boy simply standing impassively by the rope, which was all the time coiled on the floor."[31]

The next anecdote is one of my favorites. It happened in 1831, and the narrator is D.D. Mitchell who, at the time, was the primary federal representative with the tribes of the Northern Plains. During that year he led an expedition up the Missouri. When his people lost their horses, he was advised by his guide to request the hospitality of an Arickara tribe living in a village nearby. The tribe was friendly, and the prospectors were given food and lodging. A few days later they were offered to view a ceremony. An account of what had transpired was published in *Southern Literary Messenger* of 1835 and the article was titled "Extraordinary Indian Feats of Legerdemain."[32]

The actors (if I may so call them) were all painted in the most grotesque manner imaginable, blending so completely the ludicrous and frightful in their appearance that the spectator might be said to be somewhat undecided whether to laugh or to shudder. [...] [The medicine men then asked a boy to bring some clay from the nearby river.] The young man soon returned with the clay, and each of these human bears immediately commenced the process of moulding a number of little images exactly resembling buffaloes, men and horses, bows and arrows. When they had completed nine of each variety, the miniature buffaloes were all placed together in a line, and the little clay hunters mounted on their horses, and holding their bows and arrows in their hands, were stationed about three feet from them in a parallel line. I must confess that at this part of the ceremony I felt very much inclined to be merry, especially when I observed what appeared to me the ludicrous solemnity with which it was performed. But my ridicule was changed into astonishment, and even awe, by what speedily followed. When the buffaloes and horsemen were properly arranged one of the jugglers thus addressed the little clay men, or hunters: "My children, I know you are hungry; it has been a long time since you have been out hunting. Exert yourselves today. Try to kill as many as you can. Here are white people present who will laugh at you if you don't kill. Go! Don't you see that the buffaloes have already got the scent of you and have started?" Conceive, if possible, our amazement when the speaker's last words escaped his lips, at seeing the little images start off at full speed, followed by the Lilliputian horsemen, who with their bows of clay and arrows of straw, actually pierced the sides of the fleeing buffaloes at the distance of three feet. Several of the little animals soon fell, apparently dead — but two of them ran round the circumference of the circles (a distance of fifteen or twenty feet), and before they finally fell, one had three and the other five arrows transfixed at his side. When the buffaloes were all dead, the man who first addressed the hunters spoke to them again, and ordered them to ride into the fire [at the center of the apartment /tent], and on receiving this cruel order, the gallant horsemen, without exhibiting the least symptoms of fear or reluctance, rode forward at a break trot until they had reached the fire. The horses were stopped and drew back, when the Indian cried in an angry tone, "why don't you ride in?" The riders now commenced beating their horses with their blows, and soon succeeded in urging them into the flames, where horses and riders both tumbled down, and for a time lay baking on the coals. [...] I paid the strictest attention during the whole ceremony, in order to discover, if possible, the mode by which this extraordinary deception was practiced, but all my vigilance was of no avail. The juggles themselves sat motionless during the performance, and the nearest was not within six feet.

Very cute story indeed! And how reminiscent in grotesqueness of the Indian Rope trick! Who or what superimposed this cartoonish reality over the minds of the skeptical white men? Well, let's ask a medicine man of that time. The source of the following dialogue is the Protestant Missionary, William M. Johnson. In the year 1840 he visited the dying Wau-chus-co, a noted Ches-a-kee medicine man, who by the end of his life converted to Christianity.[33]

> *I said to him: "Ne-me-tho-miss (my grandfather), you are now very old and feeble, and cannot expect to live many days; now tell me the truth. Who was it who moved your Ches-a-kee lodge, and who was it who spoke when you were practicing your art there?"*
>
> *A pause ensued, when he replied: "Nasis [my grandson], you being in part of my nation I will tell all the truth; I know I must soon die. You must know that I fasted ten days when I was a young man, in compliance with the custom of my tribe; and while my body was feeble from long fasting, my thinking mind, soul, or spirit increased in power. It appeared to embrace a vast extent of country in its vision. While I was thus thinking, animals, some of frightful shape and size, monstrous snakes, serpents and birds of great variety appeared and addressed me in human language, proposing to be my guardian spirits. Whilst my thinking mind embraced these various moving forms, a superior intelligence directed me to select one of the bird-species spirits resembling the kite in looks and form. This spirit conversed with me, and told me to call upon him in time of need and he would aid me. Soon after my grandfather brought me food. I arose and did eat.*
>
> *Frequently I have seen a bright light at the opening at the top of the lodge, and strange faces were visible to me. The words of the spirits were audible to the spectators outside, but none could understand them but me.*
>
> *Nasis, I am now a praying Christian, and my days on earth are few! I have told you the truth. I possessed a power which I cannot explain or describe to you. I never attempted to move the lodge. I held communications with supernatural beings, or thinking minds, or spirits which acted upon my mind, or souls, and revealed to me such knowledge as I have described to you.*

There is a law, based on the principle *nemo moritimus preasumitur mentiri* — "a dying person is not presumed to lie," that apparently is in effect even today.

Some half a century later the clay human made an appearance on the other side of this planet.[34]

> *Here he ["Emil", the initiate] paused for a moment and held out his hand and almost immediately there was in it a large piece of plastic substance that resembled clay. This he placed on the table and began molding it into a form which afterwards took on the image of a beautiful human figure about six inches in height. So deftly did he work that the image was finished in a very short time. After it was finished, he held it in both hands for a moment; then he held it up and breathed upon it and it became animated. He held it in his hands for a moment longer, then placed it on the table, where it moved about. He acted so much like a human being that we never asked a question, but stood with our mouths and eyes wide open and stared.*

The source of this story is Baird T. Spalding's *Life and Teaching of the Masters of the Far East*, Volume 2. Spalding's books[35] contain accounts of some outrageously extravagant — hence true — manifestations of the powers of Indian initiates.

Have we exhausted all cases of animated clay personalities with the above two examples? Not by far, according to the Greek philosopher Aristophanes (5th century BC): "Humankind [...], weak creatures of clay!"

We close this section with the following eerie case. It was researched by Linda S. Godfrey, and the event is described in her recent book *Monsters Among Us*. The episode took place in a Baptist church, during a Sunday service in 1992. The account was received by this author in 2015, and, at that time, the witnesses were a "pleasant religious couple in their sixties," the male providing very detailed and artistically competent sketches (page 61 in her book). Among some 225 people attending the service that Sunday there was a dark-haired woman with ordinary features:

> [The pastor just ended his sermon and walked to sit with his family. All of a sudden the woman] stood up, let out a bloodcurdling scream, and began to [...] contort her head and body. [She suddenly] transformed into [...] a beastly creature [...]. It stood on hind legs and roared a roar that would have made a lion cower. It had fur and legs like [...] Pan, long teeth, and very long claws.[36]

Eventually the ushers and the pastor subdued the beast, and it immediately returned to a woman-like appearance.

Linda Godfrey traced other people who were attending the service that Sunday in 1992. The pastor refused to discuss the matter, while some others claimed that nothing unusual had happened that day in the church. Godfrey then attributes these responses to the reluctance of the people to evoke such a demonic event. But was it so? Or was it the case of split perception, as in the case of the Indian Rope Trick, where some people see extraordinary events, others see mundane scenes? I would guess the latter!

The most amazing hallucination scene I have read about seems to be the great chorus of apparitions and humans described in Spalding's *Life and Teaching of the Masters of the Far East,* Volume 1, page 116:

> That evening, after we had finished with our notes, we were invited to go directly to the lodge for dinner. When we arrived we found about three hundred people — men, women, and children — assembled and seated at long banquet tables.
> After we had been seated for perhaps twenty minutes, there was a deep stillness and in a moment a pale light flooded the

room. The light grew stronger and stronger until all the room was aglow and everything in the room sparkled as if thousands of incandescent lamps had been cunningly hidden and turned on gradually until all were fully lighted. We were to learn afterwards that there were no electric lights in the village. After the light came on, the stillness lasted for about fifteen minutes, then all of a sudden, a mist seemed to gather and there was the same gentle swish like the sound of wings that we had heard the evening before when Emil's [one of Spalding's guides and an initiated master] mother appeared before us. The mist cleared and standing in the room at different points were Emil's mother and eleven others; nine men and three women.

Words fail to describe the radiant beauty of that scene. When I say that, although they had no wings, they appeared like a troop of angels, I am not exaggerating. They stood for an instant as if transfixed. All bowed their heads and waited. In a moment there came music from unseen voices. I had heard of heavenly voices but I had never experienced them until that night. [...] When the twelve were seated in their respective places the same mist appeared again and when it cleared there stood twelve more. This time there were eleven men and one woman and among these was our friend of the records. As they stood there for a moment another song burst forth. When the song was nearly ended the twelve walked to their respective places without the slightest noise.

They were no sooner seated than the haze again filled the room. When it had cleared there were 13 standing, this time across the far end of the hall, six men and seven women; three men and three women on each side of the woman in the center. The center one appeared to be a beautiful girl in her teens. We had thought every woman that appeared was very beautiful but this one surpassed them all. They stood with bowed heads for a moment and the music again burst forth. [...] Then the choir of voices began. We arose to our feet. [...] All was a joyous, free

burst of music that came from the soul and touched the soul...
[...]
After all were seated, the silence was maintained for a time;
then every voice in the room burst forth in a glad, free chant
led by the 37 that had appeared.

The main feature distinguishing this amazing instance of massive interaction between ostensible apparitions and normal humans from the many cases of apparitions that we have encountered so far is mainly in the sheer quantity of apparitions at the same time and place. So, however unlikely it seems to be, postulating the veracity of this grand chorus is an immediate corollary of the authenticity and reality of common, simple apparitions. This note, suitably modified, applies throughout our narrative. Simple and brief hallucinations with veridical elements — and we have narrated many of them — lead through a chain of small steps to the veracity of the extraordinary "hallucinations" as described in this chapter.

5.6 Reality, Step 5: Perception Defines Matter

Sometime in the early 1970s, the biologist, botanist, zoologist, anthropologist, and adventurer Lyall Watson (1939–2008) was washed ashore on a sparsely inhibited island in the Indonesian archipelago, where he stayed for one year. On the first day he was greeted by the local people, and a young girl danced for him "a welcome and a history of her people." The girl was Tia; she was 12 years old. She, as it turned out, was an angelically beautiful soul endowed with extraordinary supernatural talents. A large portion of Watson's book, *Gifts of Unknown Things*, is devoted to Tia. Here we are interested in the following episode:

As my eyes adapted to the sudden leafy gloom, I realized I was
not alone there. At the back of the shadowy nave lay a fallen
tree, and on the trunk sat Tia with a little girl.

[...] Tia was talking to her and the child listened solemnly. [...] I couldn't hear their words, but Tia seemed to be having some difficulty making her point. I watched her stop and search the air for inspiration. Then she stood up [...] [and] began to dance.

[...] I found myself looking round at the trees with new pleasure. [...] Once she had succeeded in making us conscious of the trees the form of the dance changed.

[...] Then she did something impossible to describe. [...] The trees existed, Tia existed, and somehow there was a vital connection between them. [...] Setting up trees, putting herself in the picture, creating an elegant notion of the idea of the trees in her mind, and equating this with the reality of the trees around her. Making us look backwards and forwards between trees around and trees within, allowing us to establish that there was no difference.

Then she did a terrible thing.

She blotted out the image in her mind, and the other trees vanished with it. [...] One moment Tia danced in a grove shady kenari; the next she was standing alone in the hard, bright light of the sun.

My head reeled, and I blinked and rubbed my eyes until the grove began to grow at the edge of my vision [...].

[...] The little girl leaped to her feet and rushed around touching the trees, laughing gaily, and then stopped in front of Tia.

[...] Then she covered her eyes with her hands, taking the world away.

She opened up again, and there it was.

On, off; on, off. [37]

It ended with the little girl and Tia linking hands and dancing together. What a beautiful and profound case! And how exquisitely it represents the phenomenon called "reality"!

Lyall Watson got it! He concludes: "We begin to realize that our universe is in a sense brought into being by the participation of those involved in it!"

Indeed!

The very few who are initiated into the secrets of perception can induce hallucinations at will, affecting many people. According to Baird Spalding, such fellows were once called fakirs, magicians, and hypnotists. Indian Rope Trick may be a case in point: the fakirs seemed to superimpose an alternative reality directly into the minds of the spectators.

An even more delightful episode is given in Spalding's *Life and Teaching of the Masters of the Far East*, Volume 2. When a troop of bandits attacked the village where Spalding's company stayed, Jast, one of the guides and initiates, announced with calm indifference that he would neither resist nor surrender. Quoting from Spalding's book:

Then the whole band came toward us at full gallop. I confess that I was badly frightened but almost instantly we seemed to be surrounded by a number of shadowy forms on horseback galloping around us. Then these forms became more lifelike and increased in numbers. The horses [of the bandits] began plunging right and left, beyond all control of the riders, and this ended in a wild retreat with our phantom horsemen in close pursuit.[38]

The ability of the masters to induce hallucinating scenes and superimpose them over *normal* reality is also noted in Edward Abdill's *Masters of Wisdom; The Mahatmas, Their Letters, and the Path:*

Although Masters are free of personal karma, they are not free of universal karma. Should a war approach their peaceful precincts, they would have to act to protect themselves. For example, they might set up an illusory scene of an impassible canyon, thus stopping the advancing army from approaching them.

Human participation in, and interaction with, what we call material reality was not always reduced to the fictitious role of an "objective observer,"[39] as is prevalent today. That the material world depends on our perception was well known to certain old civilizations throughout our history. For example, human perception of the *material* reality of the Indian civilization some 1500 or more years ago was different compared to today's world. Consider the following quote from the Sanskrit edition of *The Madhyamaka-hrdaya-karika*, called *Tarkajvala*, usually ascribed to the sixth-century Buddhist monk *Bhavaviveka*, but possibly much older.[40]

The earth has no own-being and solidity. If yogins are capable of changing the solidity of the earth, then how is it possible to say that the own-being and of [the earth] has solidity.
The yogins traverse the water, unfathomable and vast, just as going overland. They can also turn the blazing forest into the coolness of frost and pearl. Or, after smothering a mountain top (into dust), hold them[selves] still on the top of his finger against the blowing winds of the height. Now, how could the yogins do these things (if there are the intrinsic natures in the elements)? Therefore, there is no intrinsic nature whatsoever.[41]

What is most interesting here and elsewhere is that these ancient philosophers took the supernatural abilities of yogins (or yogis) not as something that needed a proof, but as something so evident that it can be used as a reference point to draw conclusions about the material world.

Which parameters determine our perceptual input? How do we perceive the "material world"? What dominates: the eyes or the mind? Do we hallucinate all the time? Here is the answer Arten gave to Gary R. Renard:

It's always the mind that is seeing. It's always your mind that is hearing and feeling [...]. There is no exception to this. The body itself is just a part of your projection.[42]

Arten, by the way, was *hallucinated* by Renard! How ironic: a hallucination declares that the hallucinator is also a hallucination!

There is a consensual reality, and there are contested realities. The first one is an anthropocentric concept; it is an agreement on a subconscious level that a certain configuration of objects would commonly be accepted as the real world. The reality of other "living" beings may be different, even though it is shared with humans. As the wise entity Seth once said, a tree from the point of view of a snail *is* different, not merely looks different compared to how a human perceives it. When the human-perceived reality is incompatible in some way with the consensual reality, then it is contested. At the borderland of our perception, reality fails to be relatively universal and it fails to be unique. In the case of contested realities the declared winner is, by default, the perception of reality that is the closest to the normalized, materialistic paradigm. For example, is there any doubt which of the two accounts — the woman did transform into a werewolf and she didn't — would usually be deemed authentic, or even sane?

Let's turn to experimental physics and consider a single experiment and its long lasting variants extending from the year 1801, when Thomas Young performed it for the first time, all the way to the twenty-first century: the double-slit experiment. Photons or small particles of matter (say, electrons) are shot, one at a time, in the direction of a plate pierced by two parallel slits. Each photon passes randomly through exactly one of the slits, and then it is detected. *How* the detection is done changes the behavior of the photon at the time when it is passing through the slits. If a photographic plate is used, then it registers an interference pattern typical of a wave that passed through both slits (even though each photon is aimed at exactly one of the two slits). In this case the experimenter has no way of knowing through which slit has the photon passed. If the photons are observed by two telescopes, one for each slit, then exactly one of the devices registers the passing of a single photon through one of the slits, and the photons behave like particles. It is the experimenter who decides how the photon will *objectively* manifest in this world, simply by choosing the mode of observation! If he or she chooses telescopes, the photon behaves like a particle; if a photographic plate is chosen, then the photon behaves like a wave passing through both slits.

If the mode of observation affects the micro reality of small particles and whether they will manifest as waves or "matter," then the same is true for the macro world. Tia's mode of observation was different from the normal, and the *objective* reality changed.

What if the experimenter does not know at the time of the double-slit experiment how the observation has been done (telescopes or photographic plates)? The answer to that question, obtained in 2006, will be given in the next chapter, where we will consider hallucinations and time.

The following postulates summarize the data from this chapter:

- Our perception of the (material) reality affects other people's perceptions of the same reality.

- Our perception and our thoughts (can) change the (material) reality.
- Objective reality, in the form we normally perceive it, doesn't exist.

In the following chapters we will provide more support for these claims.

Chapter 6

The Illusion of Time and Space

NOW is the only acceptable time. You are in eternity NOW.

Baird T. Spalding

Illusion is commonly defined as misinterpretation of phenomena, where the reference point is the supposedly objective material world. Since our thesis is that the material world itself is an illusion with respect to the Supreme Consciousness, the notion of illusion is equated with the notion of hallucination: we perceive things that objectively do not exist. Time and space are prime examples.

6.1 The Nonexistence of the Arrow of Time

The simplest representation of time is an oriented line. The line in this model accounts for one dimensionality of time; the orientation accounts for the "arrow of time," going from the past, to the present, to the future. In this and the following section we will dispatch this arrow of time and the associated causal principle (past actions generate and cause future phenomena).

There is huge anecdotal and significant scientific evidence contradicting the causality principle. Let's start with precognition, or foreknowledge of a future event. Firstly, it is within our working definition of hallucination — if you see or perceive a future event, take it seriously, and report the experience to your physician, then you will likely be suspected of having lost your sanity. Secondly, it does contradict the causality principle and the associated arrow of time, for the perception of a genuine future event influences the present actions of the perceiver. As a somewhat extreme example, consider the following hypothetical scenario: you experience a vision of yourself killing an innocent man the day after tomorrow, and so, in order to prevent that from happening, you kill yourself tomorrow. Since the future event that was the cause for your suicide would then vanish from our linear time, there is an easy way out of this time-loop — it was the suicide of an insane person. Nobody in his or her right mind would consider the possibility of time-loops and parallel times in such cases. Nobody? Well, perhaps not — as we will see, there are scientific experiments giving rise to precisely such scenarios!

However, before we go there, let's say a few more words regarding precognition, and entertain ourselves with some intriguing anecdotes. Precognition cases often come packaged with other paranormal phenomena: NDE, OBE, channeling, and other spirit communications. There are also many cases of dream-precognitions but since dreams are not hallucinations, we will mainly stay away from this topic. There are also cases of precognition induced by meditation or through some sort of spiritual initialization.

There are special cases of precognition. Foreseeing tragic future events is called *premonition*; *presentiment* is an ineffable feeling of impending future events.

We note that precognitions are rather ubiquitous events, and that such cases have been recorded virtually throughout our entire recorded history. For example, we have a lament by the ancient Greek philosopher Aristotle, who remarked in his treatise *On Divination* that it is difficult either to ignore or to believe the evidence of precognition.

We have already encountered many cases of precognition, and, in particular, premonition. These cases usually came as a blend of several "paranormal" phenomena, including the apparent interventions of disembodied entities or spirits, which are often forebearers of gloomy future events. Recall, for example, the two cases where auditory hallucinations were heard by people warning them of impending danger (Section 2.3). The following case of the same sort allows for a "normal" explanation.[1]

> *A young woman with a rather uncommon given name was walking down a street with her husband one day when both heard her name, "Mina," shouted loudly. They stopped, [and] turned abruptly. Just then came a crash ahead in the direction they had been going. A huge cement block had fallen from the top of a building in their path. But no one who knew her name was in sight.*

The mundane explanation (that someone who knew her intervened) cannot be completely excluded. Otherwise, the fact that the warning was heard by both people all but exclude other mundane explanations.

The following auditory "hallucination" is more complicated. It was originally reported by the Italian press in October 1961:

> *A student named Luigi D., aged 18, was awoken at five in the morning by faint but insistent cries for help. He got up reluctantly, walked about the house but found nothing unusual. He would have gone back to bed, but an inexplicable sense of anxiety prompted him, instead, to lie down on a sofa in the next room. Suddenly there was a crash: the ceiling in his room had collapsed, burying the bed and furniture under a heap of rubble.[2]*

Where did the "faint" cries warning him of the impending danger come from? Was it an intervening *spirit* or was it a case of accessing an alternative reality in which Luigi was buried under the rubble?

Precognitions during hypnotic trances are especially dramatic. The following comes from a publication by Dr Alphonse Teste; the episode happened in 1841.[3]

The subject, Mme. M. was a patient of Dr. Teste's, who [the patient] had often proved to be clairvoyant. On a certain Friday, being alone with her and her husband, Dr. Teste hypnotized her and tried to find out how far she could foretell the future. She proceeded to tell them what would happen, but only in relation to herself. She said that on the following Tuesday, between 3 and 3:30 something — she could not tell what and where — would frighten her; she would have a fall, which would cause a miscarriage; at 3.30 she would faint for eight minutes, and would be very ill for the rest of the day and night. On Wednesday there would be considerable hemorrhage; on Thursday she would be much better, and would get up at 5:30 P.M. the hemorrhage would come on again and be followed by delirium; she would then have a good night, but would lose her reason on Friday evening. This state would last for three days, after which she would recover. In answer to questions, she declared that no precautions could possibly avert the accident. On awakening she ignored, as usual, everything that had happened in her trance. Dr. Teste impressed on her anxious husband the necessity of keeping her in complete ignorance of the prediction, and himself took careful notes of all the details of it, which he showed the next day to a medical friend of his, Dr. Latour. On Tuesday he [Dr. Teste] came to Mme. M.'s house, and found her lunching with her husband, apparently in the best of health and spirits. He said that he wished to spend the day with them. Soon afterwards he hypnotized her for a few minutes, when she repeated her predictions exactly, saying as before that she could not tell what would frighten her, nor where it would be. On waking she again knew nothing of her prediction. Dr. Teste and her husband determined to take all possible precautions against the accident, and, as the hour approached, not to let her out of their sight for a moment. Nevertheless, a few minutes after 3:30 she went out of the room, accompanied by her husband, suddenly saw a rat — an animal to which she had an intense antipathy; the shock of seeing it caused her to fall, and the results of the accident followed exactly as she had predicted.

Explaining this event away as a sequence of coincidences is the only refuge of the sceptics.

Humans are not the only species on our planet who can and do foresee future events. Anecdotal and scientific evidence supports the claim that, for example, dogs have a pronounced sensitivity of future events. Here is an illustrative case from *Fema*, June 1969. A man falls asleep in his chair with his dog by his legs:

[I was] aroused by Dick [the dog] pulling my trousers and making strange noises. He barked, growled, whined, and continually ran back and forth to the door. His behavior was so strange that I went to the door and stepped out on the porch. Dick indicated quite insistently that I follow him onto the lawn, and I did. It was a beautiful summer night. [...] Dick was standing at my side shivering and whining when suddenly a blinding flash of lighting and a loud crash shook the ground. I was stunned for a moment but then realized Dick was running in circles, yipping and barking and licking my hand.[4]

The chimney had fallen in the house, making a large gap in the roof, and smashing the living room with the chair he was sleeping in a few moments before. This appears to have been more that animal premonition — it is an animal *knowing* the near future events!

Rupert Sheldrake undertook some experiments concerning precognition manifested by dogs. In one of them he made thirty videotapes of a dog waiting for his owner to return, where the times of the return were chosen randomly. He analyzed the tapes and the frequency of the dog being by the windows (where he usually waited for the owner) and got a probability of less than 10^{-6} (one in one million) against the chance.

Particularly relevant for our purposes are mass premonitions of impending large-scale tragedies. There are, for example, well documented multiple cases of premonition related to the San Francisco earthquake in 1906. Another disaster with many documented cases of premonition happened in Aberfan in 1966, when a heap of coal ash slid down and engulfed a school, killing 128 children and 16 adults. (Other internet sources report 116 child victims and 28 adults.) There are many premonitions regarding the sinking of the Titanic liner on 14 April 1912. We quote from Leo Talamonti's *Forbidden Universe* (the original source is an article published by Ian Stevenson — whom we have mentioned several times in Section 4.3 — in the *Journal of the American S. P. R.*, No. 4, October, 1956).[5]

Another celebrated [sic] disaster was that of the S.S. Titanic, which collided with an iceberg and sank on her maiden voyage on 14 April 1912. Four days earlier a Mrs. Marshall, whose home was in Solent, went to the top of the house with her husband and some relatives to watch the "unsinkable" liner sail by. Suddenly she began to scream, declaring that the ship would sink before reaching its destination and that she could see hundreds of people jumping into the icy water. They tried to calm her, but she went on crying: "Don't stand looking at me, do something! Are going to let them all drown?" On the same day [the clairvoyant] Vincent N. Turvey [...] predicted that "a big liner would sink," and on 13 April he wrote to a

Mrs. De Steiger confirming his prediction and saying that it would come true within two days.

There are many other cases of premonition regarding the Titanic's sinking; covering them would take us too far in that direction, so we will finish that topic off here.

Some talented people can foresee future events during altered states of consciousness. We have the case of the British medium, Stella C., who in 1923 was investigated by a group of researchers working for the London Society of Psychical Research.[6] She was in a deep trance when she stated that she could see the front page of the London *Daily Mail* as it would be 37 days later! She gave a detailed description of the page, including the claim that the front page would contain a photo of someone, accompanied by the name "Andrews Salt" in large letters. Sure enough, when the *Daily Mail* appeared 37 days later, on May 19, 1923, its frontpage matched Stella's description, and, in particular, contained an ad for Andrews Liver Salts. It then transpired that the newspaper had not even decided to run the add until 16 days *after* Stella's pronouncement. If taken seriously, which we should, this amazing case contradicts just about everything associated with classical models of time.

There are people, mostly originating from now extinct societies of the pre-industrial world, whose "beliefs" have not been affected by the skepticism of modern times. Their natural state of consciousness is equivalent to the altered state of consciousness of modern humans, and they perceive events or objects outside the range of perception of "average" people. We have, for example, the case of an Indigenous American woman, who was able to predict future events rather casually. A Madame de Marson was worried that her husband, who commanded a military post in Acadia, had not returned to Quebec. Quoting from A.R.G. Owen's *Psychic Mysteries of Canada*:

An Indian woman, noting her concern, told her that her husband would return at a certain day and hour with a gray hat on his head. Seeing that Madame de Marson doubted this, as well she might, the Indian woman came to her at the predicted time and led her to a point on the riverbank. Almost immediately Monsieur de Marson appeared in a canoe with a gray hat on his head, quite astonished that his time and place of arrival could have been anticipated.[7]

The following case is also but one example of many similar anecdotes. It was first published in *The Spiritual Magazine* for August 1960. The author is Rev. W. Mountford.[8]

The incident occurred in the fen district of Norfolk, where Mr. Mountford had intimate friends with whom he was staying. They were two brothers, C. and R. Coe, who had married two sisters, and they lived about a mile apart on the same country road [...]. On a clear day in March [1860], at four o'clock in the afternoon, Mr. Mountford [...] saw Robert Coe and his wife driving towards the house in an open vehicle [horse-pulled carriage]. He said to his host, "Here is your brother coming." The latter came to the window and, looking out, said, "Oh yes, here he is; and see, Robert has got Dubbin [his horse that was injured and not used for a few weeks] out at last. [...] His hostess also came and looked, and said, "I am so glad too, that my sister is with him." [...] The carriage passed the window at a gentle pace and turned round the corner of the house where they could no longer be seen. After a minute Mr. Coe went to the door and exclaimed, "Why, what can be the matter? They have gone on without calling, a thing they never did in their lives before. [...]" Five minutes later, as they were sitting wondering by the fire, the daughter of the travelers, a robust, healthy lady of about 25 years of age, entered the room pale and excited and immediately exclaimed, "Oh, aunt, I have had such a fright. Father and mother have passed me on the road without speaking. I looked up at them as they passed by, but they looked straight on and never stopped nor said a word. A quarter of an hour before, when I started to walk here, they were siting by the fire; and now, what can be the matter? [...]"

Ten minutes later, Mr. Mountford, who was again looking out of the window, again saw the same two people in the same carriage driving the same horse; and he said, "But see, here they are, coming down the road again." His host exclaimed, "No, that is impossible because there is no path by which they could get on this road, so as to be coming down again. But sure enough, here they are, and with the same horse! How in the world they could get here?" They all stood at the window and saw pass before them precisely the same appearance which they had seen before, lady and gentleman, horse and carriage. They ran to the door and at once cross-questioned the travelers. [...] The travelers said that about fifteen minutes before they were sitting at home; that when their daughter started off they had no intention of going out, but had suddenly decided to follow her.

The visions of imminent visitors were once fairly common in isolated villages in Norway — people often perceived various sounds [or phantoms] announcing visitors (*vardøger*), and took them so naturally that upon hearing such announcements they would set up a meal for the expecting person. There is a report in a Swiss Parapsychological journal, where the author (Wereide) explains the phenomenon as follows: "I think there is a valid reason for this. The inhabitants of our [Norway's] rural districts have always lived in greater isolation than the people in other nations."[9] She then tells us that this isolation was the agent bringing forth the faculty of such perception. Well, an apt observation indeed. Norway used to be a large Lily's tank (Section 5.2) once, some time ago!

6.2 The Nonexistence of the Arrow of Time: Scientific Experiments

There are, of course, strictly controlled scientific experiments statistically proving that precognition is a genuine faculty. In one of Jahn-Dunne's experiments[10] with significantly positive results, the receiver "got" images from a distant sender up to three days *before* it was selected by a computer and "sent" by the sender. The description by the receiver was then matched against four images, one of them being the image from the sender. In another batch of experiments the Hungarian researcher, Zoltan Vassy, administered painful electric shocks as the stimulus to be precognized, and, not surprisingly, obtained very strong positive results.[11]

There is also an interesting experiment by Helmuth Schmidt (1969).[12] Schmidt used radioactive strontium-90 to generate random electrons that were activating an electronic circuit with four lamps at the end. The subjects were to guess which light would light up. In 63,000 trials, they had 4.4% more hits above chance, the probability of that happening by chance being 1:500 million. Then he repeated the experiment with subjects guessing which light would not light up. After 10,672 attempts the number of hits was 7.1% more than expected, with the chance of that happening by chance being less than $1:10^{10}$ or 1 in 10 billion.

Overall, scientific experiments confirm the genuineness of precognition with a probability astronomically close to 100%. For example,[13] the simple test consisting of subjects guessing the number of objects that were randomly chosen *after* their guess has been performed nearly 2 million times between 1935 and 1987. The overall results were positive with odds against chance around 10^{-25}, or 1 in 10,000,000,000,000,000,000,000,000.

It is also interesting that there is a significant difference in precognitive tests between people who meditate and people who don't meditate. According to Dean Radin, "Meditators' brains behaved dramatically differently about 1.5 seconds before the stimuli were randomly presented, but there were no differences in the nonmeditators' brains."[14]

Helmuth Schmidt devised an interesting experiment. He used an REG (Random Event Generator) to generate audio tones in a tape that were heard either by the left or by the right ear when using headphones (50-50 chance). Nobody heard the master tape, nor the copies he made. He locked the master tape up and gave the copies to volunteers who were asked to send an intention to have more tones in their left ear. Quoting from Lynne McTaggart's *The Intention Experiment; Using Your Thoughts to Change Your Life and the World*:

> Schmidt also created control tapes by running the audio device but not asking anyone to attempt to influence the left-right clicks. As expected, the right and left clicks of the controls were distributed more or less evenly.
>
> Once the participants had finished their attempts to influence the tapes, Schmidt had his computer analyze both the student tapes and the master tapes that had been hidden away, to see if there was any deviation from the typical random pattern. In more than 20,000 trials carried out between 1971 and 1975, Schmidt discovered a significant result: on both the copies and the masters, 55 percent had more left-ear than right-ear clicks. And both sets of tapes matched perfectly.[15]

According to Schmidt, his participants had changed a tape after it had been created: "their influence had reached 'back in time' and influenced the machine's output at the moment that it was first recorded. They did not *change* the past from what it was: they *influenced the past when it was unfolding as the present so that it became* what it was"[16] (emphasis by McTaggart).

The amalgamation of the past, present and future into a single "spacious now"[17] is the main thesis in this chapter, and we will come back to Schmidt's experiments in the last section of this chapter.

We mention again the experiments with *remote viewing* covered in sections 4.1 and 4.5. There are also the amazing precognitive cases of remote viewing. Pat Price, whom we have mentioned in Section 4.5, was once put in an electrically shielded room, so that his remote viewing ability could be tested. The experiment was monitored by some senior staff of SRI (Stanford Research Institute in Menlo Park, California). At 3:10 the researcher, Russell Targ, taped himself talking about some preliminaries regarding the experiment, noting among other things that the agent would be at their randomly chosen target (somewhere in the city) at 3:30 when the experiment would start. Pat Price then announced that they did not have to wait till 3:30, and he correctly described the place that the agents would *later* choose, including some obscure details (oriental pagoda restaurant at the marina).

Another such case is noted in the book *The Mind Race: Understanding and Using Psychic Abilities by* Russell Targ and Keith Harary.[18] Researchers in SRI tested a lady who could remotely view sites that were randomly chosen and visited *after* she finished the viewing, and the results were statistically very positive.

Summarizing, the remarkably rich anecdotal evidence, in combination with the many scientifically controlled experiments, proves with virtual certainty that the arrow of time, together with the associated causality principle, are not universally acceptable hypotheses. Proper understanding of the phenomenon of *time* requires a more comprehensive theory, with postulates that are consistent with the data shown so far, as well as with the examples that follow.

6.3 Past in the Present

The most direct access to the past in the present is through hypnotic regression, inducing conditions commonly referred to as previous incarnations. We have already recounted a few such cases: recall the exquisite and very veridical regression of an American woman into a life during the reign of the Spanish Inquisition described by Linda Tarazi (Section 4.3). There is also the emotional case of "Jane Evans" who was regressed into existence as a twelfth-century Jewish woman called Rebeca, killed during pogroms in England.[19] This case is also interesting because of veridical elements. For example, Rebeca found temporary refuge from the pursuing crowd in the crypt of a church. There was no such crypt in the church at the time of the regression. However, remains of something that seemed to have been a crypt was discovered six months after the regression by a worker renovating the church.

We also mention the en masse regressions performed by Chet Snow and Helen Wambach, where a large number of people experienced past lives.[20] There is one peculiar aspect of these regressions — the subjects were requested to describe the footwear they had on them, and they consistently perceived the correct footwear — if any — of the time and the place of their previous life.

Spontaneous reincarnation cases, when the discarnate people manifest as previous personalities in the memories of living humans, have been covered in Section 4.3. In these cases, the ostensible past resides within the present. We also recounted some instances of direct interaction between past and present. Such was the case of a submergence (via a "hallucination") with a past scene involving the American writer, Edith Wharton (Section 2.4).

Now we will pay attention to psychometry. It is defined as the transfer of information concerning an inanimate object into the consciousness of the experiencer. In most, but not all cases, the psychometrist receives impressions about the life and experience of the owner of an object (or of the object itself) by holding the object. In some extraordinary cases the object was not even touched by the psychometrist, yet the information received by him or her was proven authentic. It is hardly worth mentioning that from the standpoint of unhinged skepticism based on materialism, psychometry consists of self-induced hallucinations by delusional people.

There is a major overlap between psychometry and clairvoyance. In the case of the latter, the percipient receives information from inanimate objects that pertains to the present; in the case of the former, received information permeates the entire gamut of time, including what we call past, present, and future. Since we will eventually arrive at the claim that there is no basic difference between past, present, and future, it follows that there is no essential difference between clairvoyance and psychometry. Hence, we will include a few cases of pure clairvoyance, mainly in order to substantiate the genuineness of the phenomena.

We will start with anecdotal cases, some of which are nothing short of fascinating. By the end of this section, we will consider evidence from strictly controlled scientific experiments that allow statistical analysis.

Here is a relatively simple, but funny case: "Mr. Massey, going to see a medium, Miss L. Flower, gave her the glove of one of his friends, Mr. Pigott, who was quite unknown to Miss Flower. She can say nothing but: 'Pig, Pig. How absurd! That is all I can say: Pig, Pig.'"[21]

In the following remarkable case the psychometrist, a somnambulist (sleepwalker) referred to as Marie, accessed an emotionally charged past. In this case, it was not necessary to touch or see the repository object that was wrapped in paper:

She said that the paper contained something that killed a man. A rope? No. A necktie — she then continued. This was a prisoner who had hanged himself because he had committed a murder. He killed his victim with a woodman's hatchet. Marie indicated the spot where [the hatchet] was thrown on the ground.[22]

The hatchet was found in the place that she indicated.

This brings us to yet another extraordinary psychic, the Polish-Russian engineer and initiate in oriental esoteric knowledge, Stefan Ossowiecki. First, we consider the following controlled experiment, which essentially proved that psychometry is a genuine event. Our source is the book *Fifty Years of Psychical Research* by Harry Price.[23] During a session of the Second International Congress for Psychic Research (Warsaw, 1930), Ossowiecki was given a sealed package consisting of colored opaque envelopes, in which there was a message in French, a date, and a crude drawing of a bottle and a flag. The package had been prepared and brought to Warsaw by the British researcher Eric Dingwall, who did not attend the experiment, thus virtually excluding the little miracle of telepathy as a route to explain away the following episode. Ossowiecki held the package for a short while, then correctly visualized the flag and the bottle, the colors of the envelopes, and the numerals of the date, though not in the order written. The people witnessing the experiment were thoroughly impressed, and Dr (and Baron) Schrenck-Notzing, the paranormal researcher whom we will encounter again later, rushed to the medium and declared, "Merci, merci, au nom de la science!" ("Thank you, thank you, in the name of science!").

The genuineness of the phenomenon of psychometry is not contestable; there are just too many veridical cases. For example, during a séance with the medium Mary Pepper, the 42 sitters handed in that many questions in sealed envelopes. Mrs Pepper then, "took the envelopes one by one, tore a corner off, put it in her mouth, and at once gave fluent and accurate description of the writer, events on his past life, names, etc. She asserted that she was inspired by spirits who surrounded the writers, who told her all she said."

Here is one more case. The medium was Ludwig Kahn, and the investigator was Professor Charles Richet. They were in separate rooms when Richet wrote four sentences on four papers. Quoting from Richet's book *Our Sixth Sense*:

> *I put one of the papers under a copybook, on a well-lit table; another I take in my right hand which I keep tightly closed. Another I take in my left hand. The fourth, still folded, I burn to ashes with the aid of a lighted match. I have not the faintest idea which hand holds any particular paper.*[24]

Then Kahn guessed correctly what had been written on each of the papers, including the one that was burnt.

And one more example of the same sort:

> *A party of experts, writes E. W. Cox in* What am I?, *planned to test [the medium and clairvoyant] M. Alexis [Didier]. We prepared a packet containing a single word of twelve letters and enclosed in six envelopes of thick brown paper, each of which we carefully sealed. Handing him this packet he placed it, not before his eyes which were bound with handkerchiefs and wool, but upon his forehead, and in three minutes and a half, he wrote the content correctly, imitating the very handwriting. The word was by arrangement placed in the first envelope by a friend in a distant town [...] who did not inform us what the word was; and none of us knew until the envelopes were opened.*[25]

It is hardly worth mentioning that the above three extraordinary experiments are deemed not acceptable by materialistic science; among many requirements that are not met, the results were not published in a peer-reviewed journal, and the cases were not produced under the auspices of universities or research institutes. The problem of the existence of experiments where all such requirements are met is typically dealt with differently: the whole thing is swept under the rug and thoroughly ignored. We will provide one such example by the end of this section.

Here is one more "hallucinated" time-travel:

On being handed a fragment from the tooth of an extinct mastodon, Mrs. Denton, knowing nothing of its nature, described a monstrous animal of curious shape and habits. The same specimen was given to another medium (Mrs. Cridge), and she also described prehistoric animals, their surroundings, how they disported [archaic; enjoy oneself unrestrainedly] themselves, what they ate, etc. [...] Sometimes Mrs. Denton felt curiously identified with the animals she described. On holding a fragment of enamel from a mastodon's tooth, she said: "My impression is that it is a part of some monstrous animal, probably part of a tooth. I feel like a perfect monster, with heavy legs, unwieldy [difficult to move] head, and very large body. I go down in a shallow stream to drink. I can hardly speak, my jaws are so heavy. I feel like getting down on all fours. [...] My ears are large and leathery, and I can almost fancy they flap my face as I move my head. There are some older ones than I. It seems so out of keeping to be talking with these heavy jaws. They are dark brown, as if they had been completely tanned. There is one old fellow, with large tusks, that looks very tough. I see several young ones; in fact, there's a whole herd."[26]

We will describe another encounter with mastodons later in this chapter, in an astonishing and important example of psychometry! For the time being we stay with Mrs Denton for one more case. Here is what she experienced when given a small piece of meteorite to hold: "I seem to go miles very quickly, up and up. Streams of light come from the right, a great way off... Light shining in a vast distance."[27]

Back to the spectacular mediumship of Stefan Ossowiecki — he was also given a package with a piece of a meteor, but in his case the conditions were extraordinary. In 1927, when he realized the end of his life was fast approaching, the wealthy Hungarian man, Dionizy Jonky, wrapped the package in fabric tied with strings, and he sealed the strings with wax. He wanted to test the power of the psychometry of a psychic, but to exclude the possibility of any telepathy, even residual telepathy, he stipulated that the package was to be psychometrized not earlier than seven years after his death. In 1935, the package was given (still sealed, of course) to Ossowiecki, in the presence of 50 people, mostly scientists. Here is what he *saw* upon casually entering into a state of superconsciousness: "There is something here that pulls me to other worlds ... to another planet [...]. It collides with another body ... cosmic catastrophe. Tears away [...]. Showers into small fragments. [...] These are fragments of a meteorite."[28] He also stipulated that the package contained traces of sugar. He was 100% correct. Moreover, rather than merely stating the correct answer, he described the events resulting in the meteorites! This one event — if taken seriously, which I most certainly do — proves that psychometry, and clairvoyance in general, are not based on some kind of access to telepathic waves between humans.

Now that we have a confirmation of the extraordinary psychometric or clairvoyant powers of Ossowiecki and other psychometrists, we may be more disposed to seriously consider the following timeslip, also induced through psychometry. Ossowiecki "read" an archaeological object given to him and accessed a scene in the life of some human-like beings who lived someplace between what are now Spain and Italy at a time when the geography of the region was different, many millennia ago. After reading the object indifferently for a few minutes, Ossowiecki became stuck in this time! He said, "I cannot get out of here," and proceeded to describe a couple of monkey-like pre-humans. "They sit now. He makes advances to her. Takes her breasts and pulls her to himself [...]. She jumps and sits next to him. He begins ... there are no kisses. With her hand she covers his face. [...] She sits on him equestrian style. She sits still, looking into his eyes. Now she jumps up. [...] She traces with her nose along his nose, around his face. He embraces her with his whole strength ... [...] she jumps up again ... she is washing his member with water, covers it with green grass ..." And then Ossowiecki cried out again, "I cannot return." Eventually, he did return.[29]

Apparently, Ossowiecki, or his consciousness, was fully immersed in the event of a very distant past, so much so that the past became present, packaged with all associated percepts and emotions.

By the way, Ossowiecki categorically claimed that "time and space did not exist in the superconscious state."[30] This claim is supported by NDE and OBE accounts, and it is shared by true mystics (who have attained the mystic union), as was, for example, William Blake. This is precisely the direction we are heading in our narrative!

We point out again that Ossowiecki did not merely receive information regarding the objects that he psychometrized: his consciousness literally travelled through time and entered the associated past! These objects served merely as carriers of time-space coordinates that directed his travel. Here is how he explained his journeys in time.[31]

> He [Ossowiecki] told his friends that while he did not go into trance, when he moved into his superconscious mind, this became his primary level of awareness, and the room, even his own body, faded into a kind of shadow state. When the shift occurred, it was as if he were looking at a movie running in reverse; that is, from more recent to more distant events in the past. When he chose, he could freeze this reverse action. As that happened, it was as if he were suddenly in an airplane, first seeing a sort of broad overview, and then going lower until he was approximately at a point of view where his eye would normally be if he were physically standing on the spot he had chosen in the past. At this stage the action would begin again, only now going forward as it should. [...] He could go anywhere he wanted and see anything in the scene he desired; all the while a part of his mind knew he was sitting in his study or someone's house in Warsaw.

Ossowiecki was killed by the Nazis towards the end of the Second World War, in 1945.

We finish this section with a laboratory test performed under strict conditions, proving with astronomically high probability that the phenomenon of psychometry/clairvoyance is genuine.

From the statistical point of view, the most convincing experiment, complying with all scientific tenets, was performed during the early 1960s in Czechoslovakia. The scientist in charge of the experiment was the biochemist Milan Ryzl, and the subject was a mild-mannered clerk named Pavel Štepanek.[32] Ryzl would hypnotize Štepanek who would then try to guess the color of a card put in an opaque envelope. In each case he would merely touch a corner of the envelope. There were only two colors, so the probability of a correct guess was one out of two (or 0.5). In 5000 trials Štepanek got 3611 correct answers. The odds this happened by chance, or the p-value of the event, is less than 5.3×10^{-648}, or less than 1 in

1000000000000000000000000000000000000000000000000000
0000000000000000000000000000000000000000000000000000
0000000000000000000000000000000000000000000000000000
0000000000000000000000000000000000000000000000000000
0000000000000000000000000000000000000000000000000000
0000000000000000000000000000000000000000000000000000
0000000000000000000000000000000000000000000000000000
0000000000000000000000000000000000000000000000000000
0000000000000000000000000000000000000000000000000000
0000000000000000000000000000000000000000000000000000
0000000000000000000000000000000000000000000000000000
0000000000000000000000000000000000000000000 (there are 647 zeros in this number). This is so small that it could be taken that the chance of Štepenek getting that score by simply guessing is zero! There were other experiments with Štepanek, performed and scrutinized by a number of scientists, including some from Western Europe, and there was also the testimony of a famous magician who confirmed Štepanek's feat was beyond trickery.

We have established that the phenomenon of psychometry is genuine. The consequences are staggering; it follows that the past can be fully accessed in the present. Moreover, psychometry shows that the so-called inanimate matter can retain accessible data, implying that matter possesses a kind of consciousness. The last statement is a well-known ancient thesis, as well as a thesis of many profound philosophical theories, as is, for example, the monadology by Wilfred Leibniz.

6.4 Timeslips into an Emotionally Charged Past

Timeslip happens when people travel through time and space — with their material body or without it. According to the orthodox, materialistic psychology, such claims are indications of some kind of pathological state of the experiencers: a claim of full bodily travel in time-space is deemed delusional, while perception of other realms solely through consciousness (or mind) rests upon hallucinatory experiences. We have travelled far enough along metaphysical realms that offer other, more feasible, and, I dare say, more objective vantage points.

During timeslips a person, or more than one person at the time, perceive scenes from *the* past, *the* future or from *parallel* timelines. In relatively simple cases, the experience ends with a feeling that one has observed scenes from an alternative reality. In slightly more substantial timeslips, the feeling is that one has actually entered scenes that were clearly out of the present. In extreme cases the perceiver is fully *materially* immersed within the reality associated with this non-present timeline, so that there is an interaction between the experiencer, and the entities — not always human — inhabiting the perceived reality. We will provide examples of all of the above.

We start with a few relatively straightforward simple cases.

The original source of the first case is an article by Dr James F. McHarg, published in 1978 in the *Journal of the (London) Society of Psychical Research*. In January 1951, at around 2 a.m., Miss Smith was forced to start walking for 8 miles after her car skidded into a ditch. After a while she *saw* people dressed in unusual attire, checking some bodies scattered on the ground. The people did not go straight through the field but were rather coming around it, as if there was some obstacle in the field. It transpired that it had been the site of an ancient battle, seventh century or so, during which Northumbrians suffered defeat by the Picts. At that time there was a small lake which Miss Smith saw the Picts circling around.

In another case an English couple driving on an island, perhaps by the Scottish coast, again perceived many people carrying torches. This time the man in the couple, a certain Dr White, "decided to stop and ask one of the people ahead what was the case of so much activity and illumination, When the car was about 20 yards from the [inn they were trying to reach] both lights and figures disappeared."[33] Andrew MacKenzie, in his book *Adventures in Time* speculates that the people perceived by the couple were Vikings, who often camped on the island in medieval times in preparation for mainland raids.

Interesting cases! What follows will be even more intriguing.

There was a strong thunderstorm on a summer day in 1975, when Sir Denny was paddling his kayak somewhere in the wilderness of Saskatchewan. So, he grounded his kayak to wait for the storm to pass over. While waiting, he heard Indian drums and singing. He went in the direction of the source of the sound and saw a Cree village, with a bonfire, and people and children walking around. Although there was still heavy rain, the village seemed illuminated by the sun. He decided to go there, but he first returned to hide the kayak in the bushes. When he came back, the village was still there. He tried to approach it, but some barrier struck him and he fell, losing consciousness. When he regained consciousness, the village was not there anymore. He examined the ground and could find nothing but some stones in the ground, positioned in a circle. There was grass over them, and it was clear that the structure that contained the stones had been obliterated a long time before. When he returned back from the wilderness, he learned that there had indeed been a Cree village where he had seen it, and that years ago Blackfoot Indians had attacked it and had killed all of its inhabitants.

It is rather difficult to write off this experience as medical hallucination. On the other hand, the materialistic axioms leave no other choice (except the too often used claim that the story is fictional).

The last story is not an isolated timeslip involving emotional scenes from Indigenous American lore. We provide in the next paragraph one more such case. The story was published in the *Vancouver Sun,* on 28 August 1971, earlier appearing in the *Victoria Times.* Jack Scott, his wife, and one more couple camped on the Valdez Spit in Alaska, and the following happened during the night.

It must have been around two or three in the morning when I was awakened by the sound of tom-toms. I lay in the bedroll, looking up at the reflection of the dying fire on the canvas overhead and wondering why I felt no alarm. [...] Beyond the glowing embers [in the fire] I could see a great crowd of Indians dancing in the moonlight. They appeared to be in full ceremonial dress. There was no sound from them, only the steady, muffled rhythm of the drums as they moved gracefully and sinuously in and out of the perimeter of the beach fire. It was a beautiful sight to see, so much that I felt no apprehension whatsoever, merely a delight. I watched them for several minutes, and then I leaned over my wife to awaken her. "Come and see the dancing Indians," I was going to say. At that moment Graham [his friend] rolled over on his bedroll, got up noisily and threw several great pieces of bark on the fire. The tom-toms stopped. The Indians faded away.

It subsequently turned out that their camping sight was a burial ground for Haida Indians, and that there were thousands of Haidas buried all along the beach there, many of them victims of a plague of smallpox.

We also have an interesting account of a past scene from South Africa that enfolded in 1854 or soon after, when the *Orange Free State* was established in South Africa. A "gentleman" traveling in a horse-driven carriage (cart) stopped for the night. Not being able to sleep, he took a walk. We quote from *Borderland*; in the norms of that time, the language used to describe the story is openly racist:

> *What was his surprise to find that not far [...] there was a fierce battle going on, and [he witnessed] a large Boyer laager [circular camp made of wagons] being furiously attacked by hordes of savage Kaffirs. Who seemed to swarm like ants, and were fiercely repulsed from within the laager. The flash of rifles were clearly seen, but not a sound could be heard. While he was standing transfixed at the strange sight, a more than usually fierce onslaught was made by the Kaffirs, who desperately tried to climber over the piled up branches between the wagons [...] when they were met by a volley of no uncertain aim, which laid several in the dust, whilst the rest fled with the greatest precipitation. [...] [Then] the laager, or camp, seemed to open on one side, and out came a number of armed and mounted men in full pursuit of the enemy. They passed so close to where the traveller [sic] was standing that he could distinguish the horses, and in his excited state gave then a cheer, although no sound of horses' hoofs reached his ear.*

A "standard" time-slip! What is interesting is that a local then told the traveler that other people had the same experience at that spot! [34]

If the claim that the grand apparition of the battle had been perceived by many people is taken seriously, then it is impossible to explain the event in terms compatible with the materialistic paradigm. We will encounter similar cases in the next sections.

6.5. Spontaneous Timeslip Travels

Here is a simple timeslip:

> *The incident occurred in 1920 when a woman called Mrs. Hand, from Sao Paolo in Brazil, was a young child. She went to visit her grandmother in a hillside cottage on the moors just outside Bury, Lancashire. Playing in the garden she ran outside and found that the building had changed. It was suddenly darker, as if there was less window space, and all the furniture was different and older in design. A door to the kitchen was no longer there. Naturally assuming she entered the wrong house by mistake, the girl left. As she exited the light changed. So she went back into the house and it returned to the strange, old-world formation she had just witnessed. She left to go outside a second time and then, ten minutes later, when she returned for a third peek indoors everything was back to its normal appearance. The light levels were as expected and an aunt was cooking in the kitchen as anticipated.*[35]

The following rather similar case comes from Andrew MacKenzie's book *Adventures in Time: Encounters with the Past.*[36] In 1930, a certain Dr E.G. Moon was departing from a visit to a patient and stopped to think about the medicine he just prescribed. When he looked around again, he faced a completely changed scene: his car has vanished, the path he drove along was now some muddy tracks. He saw a man dressed in a coat with many capes, short top hat and gaiters. The scene was noiseless. The man stared at Dr Moon, who, not believing the evidence of his senses, decided to return in the house. The scene then disappeared. MacKenzie learned of this case from the wife of Moon's patient; Dr Moon wanted the case to be kept a secret until his death.

A somewhat more elaborate timeslip happened in 1951: "the plumber Harry Martindale who in 1951 was fixing pipes in the cellar of a very old building in York, saw a horse followed by several weary Roman soldiers pass through one wall, cross the floor and head off out through the opposite stone edifice."[37] They were knee-deep in the floor!

Apparent encounters with Roman soldiers in modern Britain was an event that happened many times. We recount a few such cases, starting with the following medieval account; the original source is the *Archives of the Northfolk for 1603*, written by Benjamin Curtiss:

Two friends and myself were swimming across the lake [...], when strangely enough we felt our feet touch the bottom. [Then he tells us that that should not have happened for the water was about 12 feet deep where they swam.] We kept together and presently found ourselves standing in the middle of a large arena with much seats one above the other all around us. The water was gone and we were standing there dressed as Roman Officers. [...] The top of the amphitheatre was all open to the sky, and many flags of diverse colours floated in the wind from the top of the walls...[38]

This happened in Wroxham Broad, an area of open water in Norfolk. Norfolk, including the city of Norwich, will be mentioned several times in our narrative. We mention in passing that, according to dowsers, there is a major vortex in Norfolk, a singularity in the intersection of some ley lines that follow the Earth's magnetic field.

Amazingly, Wroxham Broad itself was the place of at least two more major timeslips into Roman times. The next account is taken from *The Gentleman's Gazette* from 16 April 1709; the narrator is Reverend Thomas Josiah Penson.[39]

> *... we were holding a picnic on the banks of a beautiful lake in Northfolk about eleven miles from the ancient city of Norwich, when we were suddenly and very peremptorily [commandingly] ordered away by a very undesirable looking person, whose appearance and clothes belied his refinements of natural good breeding. [...] We made go away, when suddenly we had to quickly stand aside to make passage for a long procession of regal splendor: the outstanding characters of which were a golden chariot containing a hideous looking man dressed as a Roman General, and drawn by ten white prancing stallions, about a dozen lions led in chains by stalwart Roman solders, a band of trumpeters making great noise, and another band of drummers, followed by several hundreds of long-haired, partly armored seafaring men, or sea-soldiers, all chained together. They passed quite close to us, but no-one apparently saw us. There must have been seven or eight hundred horsemen in this long procession of archers, pike-men and ballistic machines. Whither they went or from whence they came I know not, yet they vanished at the lake side. The noise of their passing was very loud and unmistakable.*

Lions! How astonishingly remarkable!

And there is one more sighting of Roman soldiers at about the same place, some 120 years later. In 1829, Lord Percival Durand was with family and friends aboard his yacht not far from the entrance to Wroxham Broad:

> *... when a curious old man appeared. [...] The weird old fellow claimed to be Flavius Mantus, the Custos Rotularum ["keeper of the rolls"]. This was an extremely senior and important rank. Durand and his party were warned that they*

were trespassing on lands under the protection of the Western Emperor Carausius. According to Durand's evidence [from his private correspondence], the strange old man was suddenly transformed into a splendidly dressed Roman officer, and the waters of the broad vanished to reveal the familiar [from the precious accounts] amphitheatre and imperial procession.[40]

The same amphitheater, the same imperial procession observed in the previous accounts, and perhaps the same Roman general? The similarities of the narratives, more than a century apart from each other, are perplexing. The most obvious conclusion is that they are based on *objective* percepts!

Back to Norwich in Norfolk, here is the strange case of Mr Fey (fictitious name), coming from Lionel and Patricia Fanthorpe's book *Mysteries and Secrets of Time,* pages 145–146:

This mystery took place in the public restrooms in Tombland just outside Norwich Cathedral, in Norfolk, in one of the most ancient and historic parts of the city. [...] There is an underground public restroom in Tombland, which, at the time of the episode, had parking spaces around it. Mr. and Mrs. Fey [...] parked in one of the spaces near the restroom so that Mr. Fey could use it. The restroom itself was a very quaint old Victorian structure, and men who use it have to descend a flight of steps — the only way in and out. [Mrs. Fey waited in the car for 10 minutes and started to get worried. The parking attendant came to check the cars, and she asked him to investigate and see if anything had happened to her husband.] He [the traffic warden] searched the three washroom cubicles — all empty. [...] It was simple enough to search. The traffic warden found no signs of Mr. Fey, no signs of struggle, no blood anywhere to suggest someone had been attacked. [...] [Mrs. Fey] was deep in thought when he [Mr. Fey] appeared at the head of the stairs, walked rather unsteadily to their car, opened the door and flopped into the driving seat. He asked for a minute to regain his composure and then he explained. "I went down and used the restroom, washed my hands, and came up again. [...] When I got to the top, you weren't here, the car wasn't here, everything looked strangely different, and there were cars of type I have never seen before gliding past on that road without making sound. I just didn't know what to do. The only thing that hadn't changed was the steel handrail that I was still holding, and the steps back down behind me. I thought hard about what was the best thing to do, and then decided to go back down again. I opened one of the cubicles, lowered the seat cover, and sat there to rest quietly and try to puzzle out what had happened. Finally I decided to come back up and see if things were still weird, or whether they were normal again. And here you were."

A very strange timeslip indeed, straight into an alternative future! Or into an alternative present or even past? If the two timelines, one leading to the *normal* present, the other to the time of the timeslip, branched at a point in the *past* equidistant from the present and the timeslip, then we may call the timeslip moment an alternative present! If the former is shorter, and if we rely solely on the elapsed years, then the timeslip moment happened in the past. More about this in the last section of this chapter.

Too bad this case seems rather weakly supported (by witnesses or anything else). However, and amazingly, there is an account of a case that was rather similar:

C.H. d'Alessio [...] was walking the streets of London one evening in 1975 when he seems to have been projected into the future. He saw strange cars floating past on cushions of air and roadways with a synthetic, silvery feel to them. [...] During the episode he felt all the traffic sounds disappear, everything became muted, time lost its hold on him and he was in some oddly calm and trance-like state [...].[41]

The following simple case seems rather important in order to understand the timeslip phenomena. The account was sent by a copper miner to Allen Hynek, and the source for Linda S. Godfrey's book *Monsters Among Us* was the blog "The Biggest Study" by Michael Swards. The location of the event is a mine near Yellowknife (Canada):

The miner encountered the mist while exploring the bottom of a fifty-foot-deep gully full of what looked like very old rocks and little else except for a tubular mist that was distinct from the rest of the atmosphere. Rather recklessly he decided to walk into it and was amazed to find himself wading into a field of three-foot-high grass. He was able to step out of the mist and saw a lake-side oasis [...] and two robed, humanoid figures watching him as they floated several feet above the grass. At this point he decided to exit, which he could manage only with great difficulty.[42]

Amazing case! Notice the mist and the fact that the entrance to the other world was underground. It is a recurring theme: the public restroom in Mr Fey's case was also underground.[43]

So was the medieval timeslip described by Tu Kuang-t'ing (858–933 AD) in his *Report Concerning the Cave Heavens and Land of Happiness in Famous Mountains*. It concerns a man who entered the "cave heavens": "After walking ten miles, he suddenly found himself in a beautiful land with a clear blue sky, [...] fragrant flowers, densely growing willows, towers the color of cinnabar, pavilions of red jade, and far-flung palaces." Apparently, he enjoyed himself for a while, and then he was led out of the cave by a strange light that danced before him. But there is a twist: when he came to his village, he met his descendants of nine generations after him!

6.6 Animals Out of Time

There are events that allow other extraordinary explanations beside timeslips. Well attested sightings of the Loch Ness "monster" are such examples; if taken seriously — which I do — it is not clear whether this phenomenon is the case of a full bodily timeslip of the monster into our timeline, or it is a repeating case of group hallucinations, or perhaps it is associated with an intervention by other sentient beings, including by those that we call aliens.

There exist numerous sightings of animals that simply do not belong to our present reality. Here is a relatively recent example — *Associated Press*, October 16, 2002:

> *ANCHORAGE — Reports of a giant winged creature seen in Southwest Alaska in recent weeks have biologists and residents puzzled.*
> *Villagers in Togiak and Manokotak say the huge bird has a wingspan of about 14-feet — the size of a small plane — and is much bigger than anything they have seen before.*
> *Scientists aren't sure what to make of the reports. No one doubts that people in the region west of Dillingham have seen a very large raptorlike bird. But biologists and other people familiar with big Alaska birds say they're skeptical it's that big.*[44]

Pterodactyls — or rather wing lizards from the order pterosaurs or similar — have been *seen* numerous times and at mutually distant locations. There were, for example, numerous such sightings in the Jiundu swamps in the north-west corner of Northern Rhodesia (modern-day Zimbabwe), near the frontier of the Belgian Congo and Angola. This was published in 1923 by Frank H. Melland, who published an account of his travels entitled *In Witchbound Africa*.[45]

> [He] tells how he kept coming upon rather vague rumours about a much-feared animal called kongamato [...]. He asked what it was. The natives told him that it was a bird, but not exactly a bird, more like a lizard with wings and skin like a bat's. [Afterwards] he learned that the beast's wing-span was between four and seven feet, that it had no feathers at all, its skin was bare and smooth, and its beak was full of teeth. [...] He showed them pictures [and] they immediately plumped for the Pterodactyl, excitedly muttering kongamato!

Here is a related account from Ivan T. Sanderson:

> [He was in water, trying to fetch a fruit–bat that they shot. George, his companion shouted, "Look out!"] And I looked. Then I let out a shout also and instantly bobbed down under the water, because coming straight at me only a few feet above the water was a black thing the size of an eagle. I had only a glimpse of its face, yet that was quite sufficient, for its lower jaw hung open and bore a semicircle of pointed white teeth set about their own width apart from each other. [Later on, in the evening, the creature came again] hurling back down the river, its teeth chattering, the air "shss-shssing" as it was cleft by the great black, dracula-like wings.[46]

Later, when he inquired with the native hunters, they, upon learning that the creature was seen nearby, ran away in the direction of their village. A fine story combining the "white man's" account with the native *myths*.

Curiously, I once hallucinated a pterodactyl-like animal with round jaws (see Chapter 9).

There are also medieval European dragon/pterodactyl/ pterosaur sightings! In 1619 the prefect, Christopher Schere, observed a flying dragon emerging from one cave and eventually disappearing into another cave across Lake Uri in Switzerland. The dragon had a long neck and tail, and the head of a serpent with teeth. Herr Schere wrote at the end of his account: "Thus I wrote to you with respect, that the existence of dragons in nature is not to be doubted any more."[47]

Back to pterodactyls. In the examples we have given so far, the sightings of pterodactyls could be explained away as medical hallucinations. Not so for the multiple encounters with this or similar "extinct" protobirds that happened in Texas in 1979. The most stunning encounter happened on 14 January (1979), to Armando Grimaldo from Raymondville, Texas. We quote from Richard Lazarus' book *Beyond the Impossible*:

Armando Grimaldo was sitting smoking a cigarette in the backyard of his mother-in-law's house on the north side of Raymondville when a creature he described as looking like "something out of hell" descended upon him. With a wing- span of some 10 to 12 feet [3 to 4 meters], a leathery, blackish- brown skin, a long-toothed beak and terrifying huge, red eyes the flying creature ripped and tore at Grimaldo with its claws, attempting to drag him into the air. When other people inside were alerted by Grimaldo's hysterical screeching they rushed outside just in time to see the creature fly off into the night. Badly lacerated and in a state of shock, the injured man was taken to the local Wallacey County Hospital.[48]

There were several other, independent sightings of the same out-of-time creature.

There are numerous other sightings of creatures that clearly do not belong to our present. It is not feasible to consider them here, for that would take us too far into the intriguing subject of cryptozoology.

6.7 Interactive Timeslips

We have now arrived at the most extraordinary timeslips, when the experiencers not only perceive scenes from other timelines, but they fully interact with the inhabitants of these realities.

We start with the relatively well known timeslip experienced by two English women, Miss Moberly and Miss Jourdain, who, on 10 August 1901, while they were casually strolling in the gardens of the Palace of Versailles, experienced a major shift in time. The two ladies apparently journeyed to the period of the French Revolution at about the time of the execution, by guillotine, of the last queen of France, Marie Antoinette. Here is a short insert from Andrew MacKenzie's *Adventures in Time*[49]; the original source is Moberly and Jourdain's book *An Adventure*, published in 1910:

It was a great relief at that moment to hear someone running up to them in breathless haste. [But when they turned, they saw no one.] [...] At almost the same moment she [Miss Moberly] suddenly perceived another man quite close to them, who had, apparently, just come over, round, or through, a rock that shut out the view at the junction of the paths. [...] He looked greatly excited and called out to them, "Mesdames, Mesdames, il ne

faut pas passer par lá." [Ladies, Ladies, do not pass here.] [...]
She instantly went toward a little bridge on the right, and
turning her head to join Miss Jourdain in thanking him found,
to her surprise, that he was not there.

The Versailles case is very interesting, not only because of the
timeslip but also because of the two-way perception: the ladies
perceived Versailles during the time of Marie Antoinette, and
the people from that world perceived the ladies! They not only
saw each other, but they briefly conversed. Their case is of full-
blown time travel.

Amazingly, there are quite a few similar and independent
timeslips happening in the Versailles' gardens at the time of the
execution of Marie Antoinette, and experienced by other people,
at other times. For example, in 1938, Elizabeth Hatton, while
strolling in the gardens, *saw* some eighteenth-century people
dressed the same way as the two ladies mentioned above had
seen them:

I was completely baffled [to see a couple of people dressed as
peasants drawing a little wooden trundle cart with logs of
wood in it] and turned to watch where they were going ... as I
watched them they seemed gradually to vanish.[50]

The next timeslip that we will cover also happened in France
and may be even more striking. In the summer of 1979, Len and
Cynthia Gisby, together with their friends the Simpsons, drove
through France to Spain. Close to the Spanish border and with
night approaching they stopped at a motel where a strangely
dressed bellboy told them that the motel was full and directed
them towards a little motel along a short track. They found the
motel, managed to get two rooms, and decided to have dinner
there:

The meals were served on heavy metal plates in a room without linen or other accepted amenities. Fellow guests were some roughly clothed men drinking beer from large tankards. [...] The [bed]rooms were amazing. There was nothing modern [...], just heavy wooden trappings, shutters instead of glass on the windows, a bathroom with plumbing out of the stone age and no pillows on the beds or locks on the door. [...] Next morning they had breakfast [...] comprising of bread, jam and disgustingly strong black coffee. There were some curiosities, notably a woman who entered dressed as if she had just come from a ball, with a long sweeping gown. [...] There were also two police officers [...] [who] wore deep blue uniforms and peaked hats [...].[51]

They asked the policemen for directions, but they could not comprehend the French word autoroute (motorway). The Gisbys and Simpsons were charged the ridiculously small amount of 19 francs; there was no problem with them paying with modern francs! A couple of weeks later, on their way back to England, they wanted to stay in the same cheap motel, but the motel was nowhere to be found! It is interesting to note that Len Gisby underwent a hypnotic regression to probe the strange event, and that he gave the same account under hypnosis.

One should not overlook the fact that in these cases the perception of a time-shifted scene was collective — both in the Versailles cases and in the case of the out-of-time inn there was more than one experiencer at the same time.

The following interactive timeslip comes from Jenny Randles' book *Time Travel; Fact, Fiction & Possibility*.[52]

John Peters from Durham told me of his remarkable experience. It was October 1967 and he was driving north through the border country towards Kelso in Scotland when he got lost.

The road petered out into a track and led through some sparse woodland. Then, thankfully, he came upon a village which in truth, was a little more than a hamlet. He parked his car and went on a brief stroll to ask for directions but saw few signs of life and no other cars. But there was a woman in an odd-looking long gray dress with a shawl draped around her shoulders. She told him where the main road was located.

When he arrived at his destination, he apologized for being late and mentioned the hamlet and the strange woman. He was told no such hamlet existed. He returned to take photos of the hamlet and prove that his friends were wrong. But the hamlet was not where it was a day before. Eventually he was told that there had once been a village, but the valley was dammed to make a reservoir, and in the process several villages were destroyed and abandoned.

We have reached past the point of no return: there is the realization that we are fully immersed in the plane of perceived material existence in which time is a tentative notion, which, together with what we experience as space, merely provides coordinates to help us navigate in the world of corporal percepts.

6.8 Time-storms

The term "time-storm" is used by Jenny Randles in her book with the same title. A typical case starts with a sighting of a misty mass, in various colors, that encircles the experiencers and causes a seemingly instantaneous shift in time and space. Randles herself uncovered a significant number of such intriguing cases.

We start with a simple case, that could easily be relegated to the set of cases that allow for "normal" explanations.

David, around 18 at the time, was with his girlfriend sitting on a bridge over a stream. Quoting David: "Suddenly everything appeared abnormally quiet. [This is a recurring sensation and Jenny Randles calls it Oz factor.] [...] I then noticed a white mist about six inches above the ground." Now Jenny Randles writes: [This mist] "swirled around them, and as it did so something stunning occurred. Time slowed down. [...] His cigarette smoke was spiraling upwards far too slowly. [...] He added that, although some minutes had passed ([it seemed like hours), his cigarette had inexplicably not burned down."[53]

We pay attention to the mist: it will be an aspect of almost all similar phenomena of dislocated time and space.

An anonymous sailor told of his adventure directly to Jenny Randles. It was October 1928, and he was on a large tanker off the coast of Florida (within the so-called Bermuda triangle); at 8 p.m. he went into the library of the ship. Quoting from Randles' book *Time Storms*:

There had been no transition: it had just got extremely quiet [...]. A little perturbed, Bill [the sailor] went out on the deck to see if the engines had failed [...]. Indeed, the vessel appeared becalmed. But to his astonishment the entire ship was deserted. [There was no sound at all; the Oz factor.] But the biggest shock of all came when he went to look over the side. [...] [Quoting Bill:] "There was a grey misty sheen around us. Everywhere I looked, the sea and the sky blended into one wall of seamless grey. [...] I could see no horizon. Nothing was moving. It was exactly as if time stood still and I was no longer a part of the world." After staring at this for some moments, Bill sat down on a walkway amidships [...]. In this way he hoped to meet someone and find out what on earth was happening. Then he heard a crewmate hurrying towards him, shouting loudly and calling for others to follow. [...] [The mist had disappeared.] Bill later learned that they feared he had fallen overboard [for they could not find him]. [...] He had been missing for over an hour.[54]

One more case from the same Randles' book:

> *Social psychologist Dr Richard Sigismonde told me how, on 19*
> *November 1980, two of his friends, Mary and Michael, were*
> *driving through Longmont, Colorado, USA when a blue glow*
> *enveloped their car accompanied by a noise like rushing air.*
> *The car radio was filled with static and the headlights rapidly*
> *failed. Then the car was sucked up off the road and they were*
> *floating, gravity free, for an unknown length of time. They*
> *were surrounded by a glowing mist [...]. Next thing they knew*
> *they were further down the road [...] and it was an hour later.*[55]

There are dozens of similar cases collected by Janny Randles. The "mist" clearly plays a major role. Randles compares it with storm-bearing clouds. That may be off the mark. The mist manifests uncanny patterns suggesting the presence of some intelligence behind these occasions. In almost all of the cases the dislocation of the experiencers in space and time is not random. If the event occurs while driving a car, the experiencers manifest a gap in their consciousness, and typically find themselves some distance along the same road, in the direction of the trip or in the opposite direction. The exceptions to this rule, as it seems, happen in order to make a point. For example, there are a couple of cases on pages 77–78 in Randles' book *Time Travel: Fact, Fiction & Possibility* where a driver and a car were "instantaneously" (missing time cases) moved in the middle of some snow or mud-covered fields with no tire tracks around. It seems as if there is an intelligent entity behind the actions of the mist. This thesis will be amplified in the next chapter.

Some of the cases of time and space dislocation following an encounter with a mist (sometimes described as fog, other times as glow) are startling. Consider the following story coming from *Flying Saucer Review*, Volume 27, No. 2, 1981.[56]

On 20 April 1981, Jorge Ramos, a representative of a chemical company, left his home in Linhares, Brazil, at 6 p.m. to drive the few miles to a meeting. He never arrived. He disappeared. His wife alerted the police, and they found the car, with the key in the ignition. Five days later he called his wife. Ramos said that he had been driving to the appointment when suddenly he saw a white glow heading towards him. Before he could react it had enveloped the car. He felt a sense of pressure on him [...]. Then he found himself in a dreamy, floating state, and the next instant he was "coming to" with his body still sore and disoriented. The car was gone. [...] [He went into a nearby pharmacy and] learned that not only was it no longer the evening of 20 April (it was in fact 25th) but that he was [...] in the town of Gioania — some six hundred miles from where his car was discovered by the police.

We also have an old case that could be easily interpreted as an example of time/space dislocation precipitated by a mist in the form of stationary clouds.[57] On 12 August 1915, a whole battalion of the British Royal Norfolk Regiment disappeared during a battle at Gallipoli with the Turks. Subsequent inquiries did not reveal anything relevant; the Turks denied having anything to do with their disappearance. This could have been one of many mundane cases of soldiers vanishing in the fog of the battle, except for the testimony half a century later by a former New Zealand sapper, Frederick Reichardt. Apparently, he witnessed the event! According to him, during the battle the soldiers of the Fifth Territorial Battalion rushed into an unusual formation of about eight loaf-shaped clouds which were lying on the ground without any movement or change of shape. The soldiers never exited the cloud formation that stayed on the ground for about an hour, after which the loaf-shaped clouds rose up in the sky leaving nothing behind them.

Amazingly we have a case when a time-space dislocation was caught by a security camera. However, as we will see, this case allows yet another unconventional interpretation.

The source in this case is the *MUFON* UFO Journal, specifically the article "An analysis of missing time captured on videotape, by John Carpenter." In a nutshell, the security cameras in a small factory in Florida caught the moment of disappearance of a worker, and then his subsequent reappearance one hour and fifty minutes later. Carpenter analyzed the videotape frame by frame. The tape shows the man being distracted by something to his left, followed by the sudden appearance of a brilliant circular light, as if someone "has shined a huge flashlight from the sky above him." The entire screen then goes white, while the other three cameras viewing other points in the factory still function normally for a brief period of time. Then there is some kind of magnetic-electrical interference, and all four cameras malfunction. This lasts a couple of seconds, after which the cameras return to normal functioning. During these two seconds the worker disappears and is nowhere to be found. An hour and fifty minutes later the camera giving the view of the place of the disappearance records the return of the worker. It was preceded by the sudden appearance of an eerie, oval "puddle of light," followed by another brief darkening of the screens, followed by a view of a second "puddle of light." The worker can now be seen in the center of "a small, self-contained light." He is kneeling and vomiting profusely.

Carpenter's conclusion following his analysis of the tape was that the event was authentic and not some kind of elaborate hoax. In his article his conclusion was supported by an expert in criminology and psychology, a UFO researcher, a computer analyst, and by another computer/photography specialist.

The implied interpretation of the event is that it is a case of alien abduction. Jenny Randles prefers the time-storm interpretation. The difference between these two explanations is often superficial and almost arbitrary. There are many cases of alien abductions and missing time that were executed with a shining brilliant huge flashlight, while, as we have seen, the time-storm disappearances and missing time cases are preceded by the appearance of the mist. It seems that at least in some cases there is an overlapping of these two phenomena.

The "mist" will play a role in the next chapter too!

6.9 What Is Time Anyway?

We start with a story.

Professor Mikra — for simplicity referred to as Mikra in what follows[58] — checked the output of a randomly generated long sequence of the numerals 1 and −1; sure enough there were, as expected, about as many 1s as −1s. He did that many times and there was not a single exception to this rule. So certain he became of the even distribution of 1s and −1s in the randomly generated sequence, that he, with macabre sense of humor, decided to put his life at stake. He connected to the sequence generator a flask with cyanide that was set to automatically release its content, if the random output contained so many more 1s than −1s; the likelihood of that event happening at random was one in ten thousand or less. He activated the generator, nothing happened with the flask — there were about as many 1s as −1s again. He repeated the experiment again, and again, and again... The output was always as expected. So, he almost forgot about the attached killer flask.

One Tuesday at 10 a.m. he repeated the experiment without checking the output. He also forgot to connect the flask to the sequence generator. So he did it again one hour later, at exactly 11 a.m. This delay, as it turned out, was crucial for the success of the new experiment that he administered two days later. On Wednesday, Mikra happened to read of an experiment by the researcher Helmut Schmidt,[59] who, as told earlier on in this chapter, used a REG (Random Event Generator) to generate audio tones in a tape, heard either by the left or by the right ear when using headphones (50-50 chance). Nobody heard the master tape. Schmidt then asked volunteers to influence the output so that they get more tones in their left ear. The results were significant: in more than 20,000 trials, there were 55% tones heard by the left ear. The probability that this happened by chance, or the p-value of this event, is less than 1 in 10^{45}. Schmidt also experimented with random sequences of 1s and -1s. In 1975, he instructed 20 volunteers to try to increase the number of 1-s in a randomly generated but unobserved sequence of 1-s and −1-s that was produced *prior* to their attempts of psychokinesis. The results were also positive, and, moreover, have been successfully replicated later. The immediate consequence is the reversal of the causal principle, so that the arrow of time is seemingly being directed from the present to the past!

Returning to our story, the next day Mikra promptly gathered 20 of his friends, most of whom were seasoned meditators, and asked them to try to increase the occurrences of 1 in the *unobserved* sequence of 1s and −1s, that was generated the previous Tuesday at 10 a.m. The experiment was successful. So, on the previous Tuesday, at 11 a.m., the flask released the cyanide, which caused the death of Mikra.

This opens a conundrum of seeming paradoxes. To start with, if Mikra died on Tuesday, there should have been no Mikra the following Thursday. So then, who conducted the experiment that killed the Tuesday Mikra? We have arrived at the old "paradox" caused by the action of killing oneself in the past. It can be easily resolved by allowing branching of time into different space-time scenarios. More about this further in this section.

The one-hour delay (releasing the cyanide Tuesday at 11 a.m. instead of Tuesday at 10 a.m. when the sequence was generated) proved to be crucial for the success (!) of the Thursday experiment. Had the flask been connected to the random sequence generator at 10 a.m. nothing would have happened. The point is that in *such a case* there would be two outwardly contradictory events happening on Tuesday at 10 a.m. — Mikra continuing to live at 10 a.m. so that he could administer the Thursday experiment, and Micra being killed by the overproduction of 1s caused by the Thursday action of the influencers of the random sequence production. During the first of these two events, when Mikra survived the production of the random sequence, although he did not check the output, he would have *known* that the distribution of the two numbers in the sequence was roughly even, and so that aspect of the generated sequence would have been *known* at that time. As noticed by Schmidt and other researchers, this would make the sequence partially observed, and thereby not mutable.

While this story is fictitious, it is fully based on Schmidt's result, and so the scenario described above is rooted in our reality. In other words, the killing of Mikra on Tuesday at 11 a.m. by the experiment conducted by Mikra the following Thursday is practically doable. So, it is possible to have the dead Mikra Tuesday, and the alive Mikra the following Thursday.

We note that Schmidt's experiment is not the only one implying the above scenario. Recall the double-slit experiment described in Section 5.6: a photon is registered as a wave or as a particle depending on the way it has been observed. According to a relatively recent result of French physicists led by Vincent Jacques,[60] the experimenter had to make no choice at all regarding the mode of observation, but rather gather the information at the end of the test, conforming physicist John Wheeler's expectation that the outcome of the experiment (particle or wave) would be affected even if the experimenter got the information (telescope or photo screen) *after* the experiment. It is then easy to get the Mikra-alive-and-dead scenario based on the delayed double-slit experiment by mimicking and slightly modifying the argument based on Schmidt's results.

One way to explain the seeming paradox is to accept that there is branching of time yielding a timeline where Mikra is dead, and another timeline where he is alive. This is the mechanistic thesis of multi-dimensionality of time. Although it is a viable theory, it is a theory with limited scope. It explains the phenomenon of Micra being simultaneously dead and alive, but it falls abysmally short in accounting for the *accessibility* of the past, or the pasts, and the future, or the futures, statistically proven through many controlled experiments, and also supported by a huge amount of related anecdotal evidence. In our narrative so far, we have covered many such experiments and anecdotal cases.

In the anecdotal cases of accessing alternative or parallel pasts it is inherently difficult to rule out the banal explanations. For example, recall the very interesting possession episode when the consciousness or the soul of the 16-year-old Hungarian girl, Iris Ferczády, was supplanted by the soul of the 40-year-old Spanish working woman, Lucia, that we covered in detail in Section 4.4. We mentioned that Lucia claimed categorically that the toreadors provoked the bulls with a lilac-coloured cape, and dispatched them with a dagger, while it is widely known that the cape was always red and that the weapon was a sword. The possibility that Lucia's past was out of our acknowledged linear past is not even mentioned in the original article; it is merely left to the reader to conclude that such claims indicated weaknesses in Lucia's testimony.

The realms associated with alternative timelines are also accessible during altered states of consciousness, including dreams. The intrepid out-of-body traveler, Robert A. Monroe, describes one such reality in his *Journeys out of the Body* (1971).[61] In this realm society is technologically less advanced: there are no electric devices at all, yet there is use of mechanical power. The buses in this world were rather bizarre; they were very wide, with seats that rose behind the driver so that all passengers could see the road.

The accessibility of future events in the present, or precognition, is even more supported by experiments and anecdotes; we have encountered quite a few of both so far. We also mention the hypnotic progressions undertaken by Hellen Wambach and Chet Snow.[62] Volunteers were hypnotized and asked to visit (in their minds) this planet in the near and far future. Different people saw different futures for a fixed year. Since it has been proven that there are authentic cases of precognition, one may interpret the different views of futures as arising from objectively different futures. This would then support the mechanistic branching times theory. We will go past this theory and propose a more comprehensive one.

So, we arrive at the claim that experimentally proven and anecdotally supported accessibility of pasts (plural!) and futures (plural!) imply that the pasts and futures somehow reside and exist within the present.

We propose a theory that is more comprehensive than the branching of time theory and that accounts for all examples in this chapter. It is based on the claim that **all outcomes of all events, at micro and macro level, *exist* within Deep Reality**, and that, in our linear time, we merely choose one of the outcomes, giving rise to the perceived reality. In other words, we put forward the thesis that **Deep Reality is a constant**. This is a gigantic claim, only partially supported in this chapter.[63]

In our narrative, *Deep Reality* corresponds to what we experience as material worlds within the domain of our corporal senses. In terms of the mystic lore, the Deep Reality is the densest plane of existence, out of seven such planes. The other six are realms beyond time and space; these are the domains of pure consciousnesses: the ultimate *Deepest Reality*. We do have some tentative knowledge about this timeless and spaceless realm, seeping into our world through contacts with matter-free consciousnesses, including discarnate entities. Such contacts are established through mediums, mystic initiates, random mystic travelers, or near-death experiencers. For example, here is what Anke Evertz experienced during her profound NDE lasting nine days continuously: "Time as we know it in a material world is completely absent — everything is happening right now and therefore all at once. As soon as you turn your attention to it, you perceive everything that ever was and ever will be, simultaneously."[64] However, this seems insufficient for drawing more definite theories. So, we mostly stay away from that important topic.[65]

Our theory that "Deep Reality is a constant" implies, for example, that all combinations of winning lottery jackpots exist within the Deep Reality (but only a few may have been realized). On a more serious note, it implies the existence of realities with different gravitational constants, different speeds of light, ... all worlds of harmony, all worlds of unending destruction, and everything in between ...

Digressing for a short paragraph, it also implies that what we call reincarnations need not be experienced in linear time. For example, the Ancient Assyrian warrior that I hallucinated one night (Chapter 8), may well be my *next* incarnation, not my previous one. The sequentially of individual consciousnesses may not be synchronized with the usual timeline; in some cases, it may go from the future to the past!

Within this Deep Reality, our linear time, going from the past, through the present, to the future, becomes illusion, an arbitrarily chosen point-thick stream in the huge ocean of times. Our *freedom* of choice is manifested in the choice of a time-stream in this ocean of times. In other words, we choose roads to take during our lives, among the infinitely many roads that exist and are known to the Supreme Being.

The stupendous largeness of this ocean of time notwithstanding, the associated theory adheres to the Occam Razor Principle — that among the theories that make the same predictions, the simpler one is better. Since Deep Reality contains all possible outcomes of all events, it is an unchanging constant. So, in that sense this associated theory is the simplest of all!

A corollary of the claim that our linear time is an illusion is the corresponding claim for space. So, the (linear) time-space package that shapes our perceptions is illusionary, and we can say that our simpleton's reality is a hallucination of a sort. The real stuff, the totality of Deep Reality, will always be outside the reach of body-based organs of perception. The primary reason for this is that humans are mini hallucinations themselves. And that is the subject of the next chapter.

Chapter 7

The Illusion of Matter

We are living in mass hypnotism.

Charles Tart

We will now start exploring some of the most outrageous "hallucinations" on file. However, we remind the reader that the ballgame has changed entirely, and that may appear outlandish and difficult to believe is but a normal phenomenon within the new paradigm that has been established. One cannot accept the genuineness of Ossowiecki's ability to witness the sexual life of protohumans who lived eons ago by holding a small piece of a fossilized bone, and, at the same time accept the status quo of the old paradigm. Ossowiecki's record of verifiable cases of psychometry strongly supports the authenticity of this non-verifiable case, so that accepting its genuineness is reasonable. Simple clairvoyance, and specifically psychometry, both proven statistically, eminently establish the possibility of a close rapport between inanimate objects and consciousness and question the very meaning of the term "inanimate." As many philosophers and mystics have postulated, everything is alive and conscious in some way, and everything is loaded with accessible data.

7.1 Cloud Busters and Cloud Makers

We start this chapter with a rather bizarre example.[1]

Rolf Alexander from Orillia, Ontario (Canada) was an adventurer who tried a great many things during his life. At times he even crossed the lines of legality — he was jailed in Florida for mail fraud, and in California for practicing medicine without a licence (he practiced healing). He was already past his prime years when he tried something entirely different. While resting one summer afternoon and casually observing the blue sky sprinkled with cumulus clouds, he tried to disintegrate a cloud. It worked! He understood immediately that cumulus clouds, whose lifespan is 15 to 20 minutes, come and go, so that he might have hit upon a cloud at the end of its existence. So, he tried the stunt again, on another cloud. It worked again. And so, Alexander discovered his talent as a cloud buster!

Eventually his exploits became more widely known and the cloud-busting experiments were performed under stricter conditions. There were witnesses who would choose the sacrificial cloud, but at the same time they would monitor control clouds. It worked without exception — only the designated cloud would disintegrate in the short time of "up to 8 minutes," while the control clouds stayed more or less intact. In one instance, in 1954, the experiment was performed in front of 50 witnesses. So, Alexander's talent appeared to have been genuine.

At times Alexander varied his feat, and in some cases, he would burn a hole within a cloud. It was a public spectacle for a while, and Independent Television filmed 11 of Alexander's demonstrations.

Then the whole affair lapsed into oblivion, except for a 2009 Hollywood spoof-movie *The Men Who Stare at Goats*, where the phenomenon was ridiculed.

Upon reading of Alexander's exploits, I decided to defy the ridicule and try it myself. And why not? It is an easy experiment; all you need is an isolated cloud in the sky and a few minutes of quietly staring at it. A dosage of optimistic, or shall we call it, naïve belief in the success of the experiment may also be necessary. That may not be sufficient: the role of the invisible entities — who we know exist — in everything that we do may be a crucial factor.

Some freewheeling dosage of imagination may also be needed. There are myriads of shapes that can be discerned or imagined from a changing cloud; I mostly looked for human and animal shapes. Then I enjoyed myself as these forms morphed one into the other, expecting all the while that the cloud would dissipate into the blue sky around it.

The whole experiment turned out to be splendidly successful! Every time I tried, the cloud obligingly vanished in no time. I have repeated the feat more than 100 times and taken brief notes of the more conclusive attempts. There are 46 items in my file, spread over a period of 14 years.

A few of the cases involve controlled conditions. A friend would choose two cumulus clouds: one would be a subject of my cloud-vanishing operation, the other, the control cloud, would be known only to this friend. In each case the cloud designated for disintegration would dissolve in a few minutes — usually in less than five minutes — while, at the same time the control cloud would continue its stroll over the sky.

In some cases, the dissipation of the chosen cloud was nothing short of spectacular — sometimes the cloud melted away in less than a minute. One case was especially striking. I was hiking alone in my favorite wilderness,[2] and at the spur of the moment I decided to dare the devil by trying the cloud-busting feat on a large cumulus cloud that covered most of the sky above me. I stopped in a clearing, invoked the help of the invisible entities that control us, focused on the cloud, and imagined it morphing into various shapes, while decreasing in size. It literary melted away in 2 minutes or so. I thanked the invisible entities who I presumed assisted me and continued my lonely walk.

In another notable instance the vanishing of a sizable cumulus cloud was accompanied by tornado-like vortices within the cloud.

Is this rapport between my mind and an ostensibly inanimate object a statistical singularity without any scientific support? The answer to that question is an emphatic NO! We will elaborate in the next section. First, we briefly mention some philosophical corroboration.

Robert Monroe, the amazing OB traveler whom we have mentioned several times, is rather explicit: he stated that clouds have consciousness. He spoke from personal experience! So did the clairvoyant, Olof Jonsson, who said: "Everything in nature has a soul in one way or another."

There are theories that postulate the existence of some kind of consciousness in inanimate matter, from the smallest units of matter to large celestial bodies. The Gaia concept, in particular, accepts the thesis that our planet Earth is a living, conscious entity. Going further in our acknowledged past, we find philosophers who claimed that the smallest units of matter are endowed with consciousness, and that this feature makes the mind-to-matter interaction possible. The German philosopher and mathematician, Wilhelm Leibniz (1646–1716), called these conscious units of matter monads, and he called "monadology" the corresponding theory. A century earlier the Italian mathematician, Gerolamo Cardano (Jerome Cardan) stated: "everything that exists in nature has a soul and is animated." The German Christin mystic, Johannes Eckhart, went further. According to him "all things are equal; they are all the same in God, and are God Himself." Going further back into ancient times, we find the following in the ancient Vedic text *Bhagavad Gita*: "Know that whenever anything, whether animate or inanimate, is produced, it is due to union of body and soul."

"There is no place where consciousness stops and the environment begins and vice versa," said the wise entity Seth, channeled by Jane Roberts.

My experiments with clouds are of dubious scientific significance; their value is mainly personal. On the other hand, taken at face value my cloud experiments have clear statistical significance. The point is that during the many hundreds of attempts, I have never once encountered a cloud that was in the process of becoming larger; they always and obediently shrunk in size and vanished. Clouds come and go. In the former case they increase in size, in the latter case their size decreases. It is reasonable to postulate that these two events happen with equal probability. Under this assumption, the chance that I have randomly encountered shrinking clouds during the more than 46 recorded cases is 1 in 2^{46}. The number 2^{46} is huge; if we endeavored to write it down using standard decimal notation with fonts of the size of the simplest atoms, then there would not be enough paper to do that, for it is estimated that there are *only* 2^{50} atoms in the Earth.

Going in the opposite direction, creating clouds, there exists an ancient precedent to human cloud influencing — rain dances! For the western man, rain dances were no more than the rituals of primitive tribes. We express strong doubt in this simplistic view, and postulate that the focused and synchronized attention of a group of people to atmospheric phenomena influences the manifestation of these phenomena. In other words, when, say, 50 people dance around a bonfire, all the while requesting rain, rain will likely come. At least such an action significantly increases the chance that rain clouds may come. As far as I have been able to find, no direct scientific attempt to check this claim has ever been made.

However, there are scientific experiments confirming that global planetary conditions are influenced by the synchronous and focused attention of many. We are referring to the Global Consciousness Project (GPC), an important scientific experiment.[3] A group of scientists led by Roger D. Nelson and Dean Radin devised a network of many computers spread all over the world. The computers simply generated truly random numbers sourced in electronic noise in resisters and alike. These numbers were gathered and analyzed at Princeton. The filed-consciousness hypothesis states that the coherent mental attention of millions of people is reflected in an increase in physical order in the environment (that includes sources of random numbers). Indeed, when the data was collected it showed significant deviation from chance during global events. For example, this happened during the funeral of Pope Paul on April 8, 2005, with odds 1:20 that the deviation was by chance. Similar data from the 2000 New Year celebrations, showed even more significant deviation, with the odds of that occurring by chance being 1:3500. Overall, they have examined 185 events of such type, and the results show "clear deviation from chance, with odds against chance 36,400 to 1." This result implies that "when millions of people become coherently focused, that the amount of *physical* coherence or order of the world also increases!"[4]

During these experiments the attention of millions of people was *not* focused on the random sequences of numbers. What happens when millions of people do focus on the random numbers? The Global Consciousness Project establishes a partial answer to this question.

We do have ample anecdotal evidence that rain dances and alike influence the weather. One example is the following, coming from the book *When the Impossible Happens* by the psychiatrist and LSD expert, Stanislav Grof. Grof tells us of the experience of Joseph Campbell,[5] who was present during a rain ceremony of a Navajo tribe in New Mexico during a severe drought. The weather was so cloudless that he felt only amusement with (what he thought) was the vain effort of the shamans. But then dark clouds started gathering on the horizon and before the ceremony was over, they were all dripping wet.

From the same source, we have the following rather entertaining example of rain-inducing. Don Jose, a 104–year–old Huichol (Mexican Indian) shaman, visited Grof in California through their mutual friend Prem Das, who was an apprentice of Don Jose. There was a 7-year drought in California at the time, and someone jokingly suggested that Don Jose perform the rain ritual. He thought about that and agreed. There were dances and chants and after that, they walked down the stone steps to the ocean.[6]

The view of the Pacific Ocean in the morning light was breathtaking and overwhelming. [...] Don Jose reached the ocean shore and stood on a flat rock, about ten feet above the water line. He placed his offering [a large shell and a rabbit tail] on the surface of the rock near his feet and started to chant. The ocean was very calm that day, but, after a few minutes of his prayer, as we all watched in astonishment, a single giant wave formed on the surface and moved rapidly toward the rock on which Don Jose stood. The massive body of water reached the rock with tremendous force, but it formed at its end a spiral crest that gently swept the offering from the rock without spraying Don Jose's feet. There was no doubt in anybody's mind that the extraordinary [shaman] had interacted with the ocean as a living being and that it responded to him by receiving the offering. [...] By that time it was literally pouring...

Here is another example from another continent; it comes from George Gurdjieff's book *Meetings with Remarkable Men*.[7] One year there was a terrible drought in the Caucus province of Kars. That summer an archimandrite arrived from Russia with a *miraculous icon* of some saint, and the next day a special service for rain was held by the clergy of Greek, Armenian and Russian churches and in the presence of almost the entire population of the region. When the solemn service was over, people started heading to their homes.

Suddenly the sky became covered with clouds, and before the people had time to reach the town there was such a downpour that everyone was drenched to the skin. In explanation of this phenomenon, as of others similar to it, one might of course use the stereotyped word "coincidence," which is the favourite word among our so-called thinking people, but it cannot be denied that this coincidence was almost too remarkable.

And we have the following from *The New York Times*, July 27, 1931:

A revival of the ancient rain dance of Northern Saskatchewan Indians, despite the ban by the government agents, is reported to have occurred recently. Fields were parched [dried up, thirsty] and cattle were suffering when Chief Buffalo Bow, head of the File Hills Reserve, decided to invoke the Great Spirit. The forty-eight-hour dance, led by six singers in relays, centered about a great tree [...]. The Great Spirit seemed to answer, for soon after the mystic rites had been performed the rain began and continued for two days, July 14 and 15, bringing relief all over Saskatchewan.

Destroying or bringing about clouds by sheer intent falls under the general scope of telekinesis. This is the subject of the next section.

7.2 Telekinesis: Mind Influences Matter

Telekinesis is standardly defined as the *supposed* ability to move objects by mental power. The attribute "supposed," almost always inserted before the word "ability," is an obligatory denigration of scientific research and of the anecdotal evidence supporting the genuineness of these phenomena; it is both an insidious brainwashing device, and a self-congratulatory statement expressing affiliation to dogmatic scientificism.

Related to telekinesis is the ability to alter properties of inanimate objects through sheer mental power. There is also a substantial body of scientific experiments supporting its genuineness.

We point out that the phenomena we will discuss in this section are the *least* extravagant compared to what we will cover in subsequent sections in this and the next chapter. We also note that telekinesis and similar phenomena are, by and large, not within the scope of medical hallucinations. We cover them here in order to offer tangible support of their authenticity, and as a gateway to the extraordinary phenomena that will be covered further in this and the following chapter.

One more thing requires clarification: the statement "mind influences matter" is not clear without addressing the colossal problem of defining "mind." We will not attempt here to properly define this monumental notion. However, we must point out that by "mind" we do not necessarily mean "human mind," commonly associated with the function of the brain. Indeed, when a talented individual moves matches, or changes the acidity of water by sheer concentration, we leave wide open the possibility that the action by this individual is indirect — that he or she acts as a conduit or medium and merely invokes powers of entities residing in *higher* planes or spaces, and that these entities actually do the work!

We have already mentioned a few experiments indicating that the mind could interact with inanimate matter. One example is the influence of focused intent on randomly generated sequences of two symbols, which we have discussed in Section 6.6.

We mention the inaugurating research by J.B. Rhine where subjects apparently influenced the outcome of a dice throw.[8] In Dean Radin's experiment, the subjects affected trajectories of photons,[9] while in William Tiller's experiment (1997) a change of pH (the measure of acidity) of water was achieved by meditation.[10]

The subjects of the above experiments were ordinary and more-or-less randomly chosen people. It is altogether another story when the experiments were performed with talented, psychic, or spiritually initiated people. Then, the results are often extraordinary in many ways. Take, for example, the talented psychic Ingo Swann, whom we have mentioned in connection to OBE (Chapter 4). In an experiment by the physicists Harold Puthoff and Russell Targ, Swann increased and decreased at will the magnetic field within a superconducting magnetic shield. In another experiment he "has been able to deflect the needle of a magnetometer encased in a superconductive substance and buried in concrete."[11]

The Russian sensitive, Nina Kulagina, caused some extraordinary phenomena, including inducing movement and, in some cases, levitation of inanimate objects at distance. There are many videos from the 1950s and 1960s showing her performing under laboratory conditions, not only moving objects, but affecting the heart rates of living creatures, including humans. The disingenuous reaction of the western world in general was to dismiss the whole affair as Soviet propaganda, implying that the Soviet Academy of Sciences, where the experiments were performed, was a corrupt institution engaged in fake investigations.

The implication of the acceptance of the genuineness of Kulagina's ability to levitate inanimate matter is massive, since with that one accepts the proposition that the mind can deny and defy gravitation.

In the next four sections we will present some amazing cases of levitation.

7.3 Levitation of Heavy Objects

We point out before we begin that most of the cases we will refer to in our narrative are relatively old. This is forced — the most outlandish manifestations of mediumistic phenomena occurred during their peak, which happened in the period between 1850 and 1935. The general population of the Western world during that period of time was still relatively free from uniformity, despite the rising scientifistic stigma a priori assigned to everything and everybody that challenged the basic materialistic premises. These were the times when "miraculous" feats produced by talented people had a wide and intelligent audience. This proliferation of mediumistic phenomena at that time eventually led to widespread institutional scientific investigations of paranormal phenomena, which peaked during the second half of the twentieth century. By the end of the twentieth century both mediumistic phenomena and their scientific investigations tapered off and it seems that they declined in the twenty-first century; the social and scientific globalization flattened the accepted and the acceptable paradigms into a simple materialistic worldview.[12]

The phenomena caused or facilitated by Kulagina mentioned in the previous section pale when compared with the spectacles occurring during séances with various mediums, especially during the second half of the nineteenth century and first quarter of the twentieth century. These are usually dismissed out of hand as not adhering to scientifically controlled conditions, despite the strict conditions often imposed over the medium and the sitters.

We start with Ossowiecki. As witnessed by many people, he caused a large marble statue weighing several hundred pounds, to move to the other end of the room. As a precaution against cheating, he was stripped, wrapped in a straitjacket and laid on the floor of a ballroom. We are told that all present, including Ossowiecki himself, were stunned by what they witnessed.[13] Another case of group hallucination?

We will now mention a few extravagant and somewhat comical cases of levitation of an object — a heavy piano!

Camille Flammarion tells us in *Mysterious Psychic Forces* about an 11-year-old child who caused, "A piano weighing more than 750 pounds [to rise] up off of its two front legs."[14]

We also have the following fantastical case of levitation of a piano.[15] During a séance with the medium, Katie Cook, the sister of the famous medium Florence Cook, a heavy cabinet piano was carried over the heads of the sitters to the opposite side of the room where it fell on the floor.

There are still more *unbelievable* occurrences of piano levitation! In the summer of 1875, a grotesquely fantastic phenomenon occurred during the mediumship of a certain Mrs Young,[16] in whose presence a heavy piano with seven (!) ladies and gentlemen on top of it, was lifted up in the air.

Even more grotesque was what happened in Florence, 1855, in the presence of the phenomenal medium D.D. Home.[17] A grand piano rose in the air, while at the same time Countess Orsini was playing on it. The piano, countess and all, remained suspended in the air during the whole piece. The same outlandish episode happened with Home himself playing the piano.

All of these is merely a prelude to even more extraordinary cases of levitation presented in the following sections. However, we first pay more attention to scientific experiments in levitation.

7.4 Levitation: Scientific Results

It seems that the object that levitated most often during séances was the table around which the sitters and the medium sat. Here is a rather typical account, described by Sofia de Morgan, the wife of the eminent mathematician, Augustus de Morgan:

*After sitting some time we were directed by the rapping to join hands and stand up round the table **without touching it**. All did so for a quarter of an hour, wondering whether anything would happen, or whether we were hoaxed by the unseen power. Just as one or two of the party talked of sitting down, the old table, which was large enough for eight or ten persons after the manner of a lodging-house, moved entirely by itself as we surrounded and followed it with our hands joined, went towards* *the [skeptical] gentleman out of the circle, and literally pushed him up to the back of the sofa, till he called out, "Hold, enough."*[18]

Now, compare Sophia's account with the following:

[During] the 11th meeting, [...] the table, instead of merely tilting or rocking on two legs, as it has done so far, rose clear from the floor. The explanation of unconscious muscle action was suddenly no longer applicable, since one cannot push the table up in the air, either consciously or unconsciously, when the hands are on top of it.

Gradually the movements [of the table] became bolder and the lamp was lit for longer periods. By its red glow we could clearly see our hands on top of the table. [...] Because the levitations were not too high, I said: "Come on — higher!" at which the table rose up chest high and remained there for eight seconds. [...] At one point the table levitated and floated right across the room: we had to leave our seats to follow it; it appeared to be about five inches off the floor, and the signal lamp remained alight until we crashed into some other furniture.

Is this the account of yet another group of spiritualistic suckers? Not at all! The above is a brief summary of 1 out of 200 scientifically controlled experiments, performed over a period of 18 months. The principal investigator was British psychologist, Kenneth James Batcheldor (1921–1988), and the results were published in the *Journal of the Society of Psychical Research* ("Report on a Case of Table Levitation and Associated Phenomena", 1966).[19] The sitters were randomly chosen volunteers. A scientifically controlled table-levitation experiment with positive results? Isn't that amazing?

There is one very interesting feature of these experiments that must be noted: 80 out of 200 times the sittings were held with a certain Mr W.G. Chick, and these were the ONLY occasions when phenomena occurred. According to Batcheldor, "positive results were almost guaranteed in the presence of W. G. C., even if there were only one more sitter [Mr Chick and himself]." So, it seemed that it was necessary to utilize the talents of at least one accidental medium to get positive results. One consequence would be that the scientific requirement of repeatability of the experiment is not guaranteed; a Mr Chick is required as a channeler from the ultimate source of these phenomena to our sensual reality.

Amazingly, about four years after Batcheldor's article was published, the results were replicated under similar conditions. C. Brooks-Smith and D.W. Hunt[20] assembled a group of four people, who knew each other well and who were interested in telekinetic phenomena. They were set up in "laboratory conditions," under full lighting and with cameras at hand.

Sitters at the table rested both hands gently on the table top, in full lighting conditions. [...] Normal conversation was carried on during the sitting but no doubt each person, in his [one of them was female] own way, was willing or wishing for the table to make a movement. Results were in fact rapidly produced: knocks and raps were heard, apparently from the table, at the first meeting. Tilting of the table occurred at the second meeting, and after a few sessions, the phenomena had developed into violent table movements over which no exact control seemed possible, and which indeed caused anxiety due to the possibility of injury. [...] These movements were produced at early stages of successive sessions and included the rising of the table some five or six feet clear of the floor, its movement over the whole of the room whilst in the air and a peculiar oscillating descent to the floor. During all of the movements the experimenters as far as possible maintained a light one-finger contact with the table, but these were unavoidably lost on occasion and many movements were possibly made without contact.

Further, from Sitting No. 3:

> *Almost at once the table became levitated and remained in a gently floating state. On one occasion when all hands were withdrawn simultaneously, the table jumped still further rather violently and hands were promptly replaced. A repetition of this procedure shortly afterwards produced moderately [severe] jumps quickly followed by the descent of the table to the floor.*

It's hardly worth mentioning that these two fascinating articles were almost completely ignored both by the scientific establishment and the public at large. Unlike the other cases, standardly dismissed as apocryphal, trickery or (medical) hallucination, scientifically controlled experiments, when they are not promptly swept under the horizon, are typically "debunked" as lacking rigor.

7.5 Levitation of Very Heavy Objects

It seems that there is no object that is too heavy to be levitated. However, the heavier the object, the more pronounced the need for a focused and intense concentration. In some special cases that is achieved by the sheer talent of the influencer. This seems to have been the case with the Brazilian medium, Mirabelli. Quoting from A. Da Silva Mello's book *Mysteries and Realities of This World and the Next*: "[Mirabelli caused] an automobile, in which several friends of his were travelling on a public road [to be] raised to a height of three meters above the ground, in which position it remained for three minutes."[21] Mirabelli's miraculous feats have been authenticated by 555 signatures, 72 of which were of doctors. The affidavit contains the following amusing statement: "You will say this is madness but we swear that it is the truth!" The truth sometimes comes wrapped in madness foil.

We now touch upon the topic of focused and coordinated action by *many* individuals, producing gravitation–defying phenomena. The original source of the following fascinating account is an article in German that appeared in 1950; an English translation is given on pages 213–215 of David Hatcher Childress' book *Anti-Gravity & the World Grid*.[22] The story concerns a Swedish doctor, Dr Jarl, who in 1939 travelled to Egypt on behalf of the English Scientific Society. From Egypt he was called by a Tibetan friend to urgently come to Tibet in order to treat a high Lama. Now quoting:

One day his friend took him to a place in the neighbourhood of the monastery and showed him a sloping meadow which was surrounded in the north west by high cliffs. In one of the rock walls, at the height of about 250 metres was a big hole which looked like the entrance to a cave. In front of the wall there was a platform [...]. The access to this platform was from the top of the cliff and the monks lowered themselves down with the help of ropes.
In the middle of the meadow, about 250 metres from the cliff, was a polished slab of rock with a bowl-like cavity in the centre. [...] A block of stone was manoeuvred into this cavity by yak oxen. The block was one metre wide and one and one-half metres long. Then 19 musical instruments were set in an arc of 90 degrees at a distance of 63 metres from the stone slab. [...] The musical instruments consisted of 13 drums and six trumpets.
[Eight of the drums were large, four of medium size, and there was one small drum. The trumpets were all of the same size.] [...] The big drums and all the trumpets were fixed on mounts which could be adjusted with staffs in the direction of the slab of stone. [...] Behind each instrument was a row of monks.

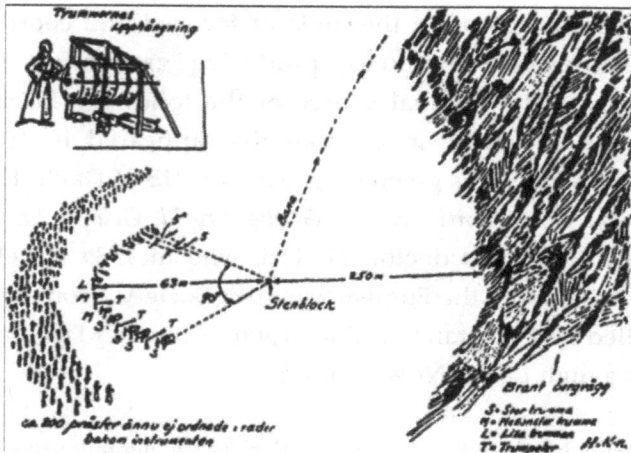

When the stone was in position the monk behind the small drum gave a signal to start the concert. [...] All the monks were singing and chanting a prayer, slowly increasing the tempo of this unbelievable noise. During the first four minutes nothing happened, then as the speed of drumming, and the noise, increased, the big stone block started to rock and sway, and suddenly took off into the air with an increasing speed in the direction of the platform [...] 250 metres high. After three minutes of ascent it landed on the platform.

Continuously they brought new blocks in the meadow, and the monks using this method, transported 5 to 6 blocks per hour on a parabolic flight track approximately 500 metres long and 250 metres high.

The usual interpretation of the minority who take this account seriously — among the very few who are familiar with this event — is that the stones were levitated by means of a kind of sonic technology. We claim that the rhythmic "music" was directed at the minds of the participating monks as a means to synchronize their mental focuses.

Dr Jarl himself initially believed that he was the victim of a hallucination. So, he made two films, and the two films corroborated what he witnessed. According to the German article, "the English Scientific Society for which Dr Jarl was working confiscated the two films and declared them classified. They will not be released until 1990." This was written in or around 1950. The two films were, of course, never released. Yet another conspiracy? Maybe. Conspiracy is another word for truth these days.

Metaphysical lore contains numerous instances of claims that various ancient structures were built precisely by the levitation "technique." Examples include the old city of Baalbek, where exceedingly heavy monoliths, the largest weighing at least 800 tones, were used to build temples; the Micronesian ancient city of Nan Madol, where according to the surviving local legends, the stones were transported by "magic"; and, of course, the grand pyramids of Giza. There are many cases of people experiencing the construction of the pyramids by the levitation of stones while being hypnotically regressed into "past lives." For example, the many books of the hypnotherapist Dolores Cannon contain fascinating transcripts of many such cases.

7.6 On Denying Gravitation

We saw earlier through psychometry that inanimate matter possesses caches of memory-like information accessible to talented people. In this chapter we have exhibited anecdotal and scientific evidence supporting the thesis that mind could interact and influence the supposedly inanimate matter, all the way to denying some basic properties of matter, such as the gravitational pull. The statement that mind could deny gravitation is a huge claim. If we could free a stone from the gravitational pull of the earth by focusing our mind on the stone, then perhaps entities with minds truly superior to our individualized consciousness — and I believe that they exist — could free the Earth from the gravitational pull of the Sun. The gravitational force of the whole universe becomes an aspect of misinterpreted reality. This misinterpretation is the home of our corporal existence.

In the following chapter we will argue that we, as defined by our bodies and our memories, are also illusions, or hallucinations.

Chapter 8

Hallucinations Are Us

Man has no body distinct from his soul, for that called
body is a portion of soul discerned by the five senses.
William Blake, *The Marriage of Heaven and Hell*

From the scientifistic point of view, experimental scientific research can be called such only if it is done under the auspices of a university or a specialized research institute. If an experiment is performed at a private residence, it does not matter how strict the conditions are, it is usually dismissed citing inadequate control of the parameters associated to the experiment. Mediumistic experiments, the success of which require special conditions not met by the inquisitorial rigidity of typical laboratories, are thereby automatically dismissed. We regard such restrictions as artificial barriers and a dogmatic buffer against anything that challenges the sanctioned axioms of materialistic science. Therefore, we reject them.

8.1 Levitation of Humans

Human bodies are *material*, and it is not surprising that the levitation phenomena apply to them too.

We begin with Louis Jacolliot. He was a colonial judge during the French possession of India in the mid-nineteenth century. These were the times when the mystic reality of ancient India was a vibrant part of everyday life, with various fakirs, holy men, yogis, masters and other initiates roaming the land. Like most Europeans, Jacolliot entered Indian reality full of skepticism. In particular he "conceived [the fakirs] to be simple magicians and [he] unceremoniously dismissed them whenever they presented themselves." His views on these matters gradually changed, especially after he met Covindasamy, a travelling fakir. During Covindasamy's third visit, Jacolliot asked him to levitate. Quoting from Jacolliot's *Occult Science in India and Among the Ancients*:

> *Leaning upon the cane with one hand, the Fakir rose gradually about two feet from the ground. His legs were crossed beneath him, and his made no change in his position, which was very like that of those bronze statues of Buddha... [...] For more than twenty minutes I tried to see how Covindasamy could thus fly in the face and eyes of all the known laws of gravity; it was entirely beyond my comprehension; the stick gave him no visible support, and there was no apparent contact between that and his body except through his right hand.*[1]

The frustration that Jacolliot showed in this case, and in other cases is comical. He studiously examined every detail, trying to find the hidden *normal* explanation, then was flabbergasted when he failed again and again to find it.

Eastern and middle eastern lands teemed with paranormal phenomena, including cases of levitation. Recall, for example, the partial levitations — or induced mass hallucinations (which may be the same thing) — in the Indian Rope Tricks covered in Chapter 5. The initiates of high order sometimes levitated at will; the great Tibetan yogi, Jetsun Milirepa (c.1052– c.1135 CE), "the most learned professor of the Science of Mind," whose initiation included 11 months of non-stop meditation, could fly! Quoting Milirepa: "Therefore, I persevered in my devotion in most joyous mood, until, finally, I actually could fly."[2] However, as he was seen flying by many people, and since "worldly fame and prosperity might retard the progress," he had to move to another, more inaccessible cave.

We should keep in mind the controlled experiment by Batcheldor and others resulting in the levitation of a table. It provides scientific support of the phenomena we cover in this section, however implausible and far-fetched they might appear to be. Thus, believing in the authenticity of these phenomena is more an expression of open-minded analysis of the whole contextual paradigm than they are an indication of credulity.

The *recorded* European cases of levitation of humans by far outnumber such recorded cases elsewhere. According to Oliver Leroy's *La Levitation*, 60 Catholic saints were seen to levitate during their lifetimes. Other sources give the number as 230 Catholic saints to whom this feat is attributed. Typically, the phenomenon occurred during exalted states of supreme detachment — rapture in cases of levitation of Christians (nirvana for Buddhists, samadhi for Hindus).

Consider, for example, the case of the Franciscan friar Giuseppe, or Joseph, of Cupertino.[3] He levitated dozens of times in the presence of many people. Among the spectators was Duke Johann Friedrich, the first patron of the philosopher and mathematician Leibniz, who converted from Protestantism to Catholicism in 1951 after witnessing Joseph's levitation on two separate occasions.

Joseph's levitations were involuntary, and he was often embarrassed by the episodes. Apparently, the mind that moved his body was external to him. The levitations at times lasted for an hour or so, and a few times they happened several times during a single day. At times Joseph flew some thirty meters high. When he levitated, he was as stiff as dead; even his clothes, including his tunic, stayed undisturbed, as if he was a still holographic image superimposed over the "normal" reality.

And then we have numerous cases of levitations by mediums during séances. It suffices to consider only one of the mediums to justify this attribute "numerous": "There are at least a hundred instances of [the medium] Mr. [D. D.] Home's rising from the ground, in the presence of as many separate persons..."[4] Particularly extraordinary was Home's levitation during a séance in December 1868:

Home, while entranced, floated out of an upper-storey [sic] (third floor) window of one room and in at another in the adjoining séance room, the windows being seven feet apart, with only a 3 inch ledge between. [...] Afterwards, Lord Adare went to the back room and found the window open about a foot, he closed it and, returning to the séance room, said he could not think how Home could have managed to squeeze through. Home then said "Come and see." "I went with him," said Lord Adare, "He told me to open the window as it was before. I did so; he told me to stand a little distance off; he then went through the open space, head first, quite rapidly, his body being nearly horizontal and apparently rigid. He came in again, feet foremost; and we returned to the other room."[5]

Amazingly there exists a relatively simple set of instructions for inducing levitation; the phenomenon is aptly called Party Levitation. It can be found, for example, in the researcher Thelma Moss' book *The Probability of the Impossible*, pages 133–136.[6] One person, the levitatee, is seated on a chair, and four other people (the levitators), two males and two females, are standing around the chair, with their hands stacked on the head of the levitatee. Now quoting from Hatcher Childress' book *Anti-Gravity & the World Grid*, page 55:

First, the levitators should be positioned 45 degrees off the magnetic compass direction of north, south, east, and west for maximum effectiveness. Second, alteration of male and female sex of the levitators adds to the gravity antenna's power. Third, the hand stack on the head of the central levitatee by the levitators should not have like-gendered (male/male, female/female) hands touching. Forth, there's no need to think of anything — just hold the hands stacked on the levitatee's head for a count to ten. On the tenth count remove the stacked hands quickly and place one finger each on the four corners of the chair. The person in charge of counting says "lift" and up goes the levitatee.7

The reader will likely be skeptical at this point. So was the publisher of Moss' book, who was reluctant to print the recipe without proof that it worked. He was so disbelieving that at first he even refused to try it himself. Eventually he relented, tried the recipe with friends, and found out that it was "simply very easy."

As we saw, at times humans are immune to gravity. This is a fact and a scientific heresy at the same time. On the other hand, hallucinating objects are always immune to gravity; such examples will be given in the next chapter. So, humans, or human bodies at times, function like hallucination. We would like to go all the way and claim, with some justification, that humans, and matter in general, are hallucinations.

We end this section with a "prank." In 1976, the British astronomer, Patrick Moore, announced on BBC Radio 2 that at 9:47 a.m. that day a once-in-a-lifetime astronomical event was going to occur: the planets Pluto, Jupiter and Earth would align at precisely that time, and that would lessen the Earth's own gravity. Moore told his listeners that if they jumped in the air at the exact moment that this planetary alignment occurred, they would experience a strange floating sensation. When 9:47 a.m. arrived, BBC2 began to receive *hundreds* of phone calls from listeners claiming to have felt the sensation. One woman even reported that she and her eleven friends had risen from their chairs and floated around the room. We postulate insolently that the joke was not on the audience, that it backfired as a joke on the astronomer himself! A genuine belief can be a powerful reality modifier — and the belief in science is the most fervent in modern times. If scientists tell us that the gravitation will certainly cease tomorrow, we will all start flying like flies!

8.2 Apports

On June 12, 2021, as I was preparing to retire for the night, I noticed the following on my pillow:

Chinese talisman: the inscription in the tablet is a blessing.

It lay there with the strings perfectly parallel, more so than in the photograph. It was some 40 centimeters (one and a quarter feet) away from the wall. So, it did not fall from the wall; someone had carefully placed it on the pillow. This could be claimed anyway, since it had never been hung on that wall or anywhere else in our house to start with. In fact, I don't recall having seen it before. If we exclude the possibility that I brought it into my bedroom by moonwalking from some unknown place and placed it on the pillow in a similar fashion, there remains the following "miraculous" scenario as the only explanation: it was an apport!

An **apport** is a material object that has been moved instantaneously from one place to another. The destination is usually a room that is far from the origin, where in many cases the origin is unknown. Apports, being normal material objects, can hardly be classified as medical hallucinations. So, the *normal* explanation of these phenomena is that they are either apocryphal, or that they are results of trickery. There are just too many instances of the apport phenomenon occurring under strictly controlled conditions that cannot be explained away in such a banal manner. And if apports are genuine, then the alleged and widely accepted thesis of objectivity of matter goes right through the window. Matter becomes hallucination. In particular, "we," meaning our bodies, become hallucinations. Corroborating this claim is the main goal in this chapter.

Apports happening seemingly randomly to people in mundane circumstances are gifts and messages. The general message is that there exist tangible phenomena that strongly indicate the existence of a reality beyond the reach of our faculties of perception and contradicting certain basic scientific laws.

Since manifestations of apports often allow mundane explanations, they are discrete and nonobtrusive. I suspect that they happen much more often than acknowledged. How many times have we searched for an object, eventually finding it mockingly displayed in a most obvious place? As the Ancient Greek philosopher Heraclitus, who was fond of semantic paradoxes, once said, "If we do not expect the unexpected, we will never find it."

There are, on the other hand, cases when banal explanations are not possible. Consider, for example, the following case. Dan, described as an "accomplished academic teaching at a prestigious institution," was making [...] blueberry muffins in his kitchen. Now quoting from Strieber–Kripal's book *The Supernatural*:

Honey like honey, he got some on the lip of the jar and so washed the jar off in the sink [closed it], and sat it down to drip dry. [...] [He then] pulled the [metal, flour] tin off the shelf, [when] it suddenly got heavier. [...] The tin dropped to the floor. [Now quoting Dan:] "Upon meeting the carpeted ground, the tin lost its lid and much of the powdery content. Rather upset at myself, I kneeled to clean up the mess. Then came the electric discovery whose current still flows through me. Enough of the flour had run out to reveal that something was buried at the bottom of the tin. Naturally curious, I dug through the flour with my fingers and then pulled out, of all things, a glass honey jar exactly like the one I had held in my hands and washed a moment ago, a jar completely caked with flour — as if it had been placed in the tin still wet. Puzzled, I turned my head to assure myself that the bottle I had just rinsed was standing where I had left it. It was not." [Now Kripal:] Dan stared for two minutes, examining the situation and its impossibility. [Dan:] "The fact was obvious. The wet jar had been moved from the sink and deposited on the bottom of the flour tin. The explanation, however, was not at all obvious." Dan considered the possibility that he had somehow hypnotized himself, but he could not replicate the phenomenon by placing the jar in the tin. His conclusion was a good one: "I knew at that instant that materialism is false."[8]

And so it goes!

To set the stage for more fantastic apport cases, we reiterate: there exist cases of apports in strictly controlled scientific experiments. For example, during Batcheldor's experiments mentioned in section 7.4, besides levitation — and to the bewilderment of both the experimenters and the subjects/sitters — there appeared apports![9] A few times small objects like matches and pebbles mysteriously "dropped" into the room from nowhere. Even large stones fell, not unlike poltergeist phenomena! One of these stones was given to a London museum for analysis. Subsequently Batcheldor received a baffling note from the museum officials saying that the rock had disappeared. How appropriate for an apport!

One of the first recorded cases happened during séances conducted by the mathematician, astrologer, psychic researcher, spy (and what not) John Dee (1527–1608), the original prototype for the James Bond movie character. During Dee's experiments with the medium Edward Kelly, and to the great delight of Dee, the manifesting angel, Michael, apported a ring with a seal. We read from Dee's first book of mysteries: "Then he [angel Michael] toke [took] a ring out of the flame of his sword. [...] After that, he threw the ring on the [...] table."[10]

Let us now recount some of the more interesting cases of apports. The following intriguing case is taken from Alexander Cannon's book *The Powers That Be* (1935):

During the spring of 1929, Medicine Capitaine Dubois was residing [in the village of Marhoum in Algeria], fighting a diphtheria epidemic. He was a perfect sceptic and was one day told of a certain Shepard named Abdul Ouab who was possessed of supernatural powers. [...] Ouab was summoned and asked to perform the "trick." He told the captain to think of some object in his palatial house in Paris, which he would like to see. Dubois concentrated his mind on a very valuable family portrait valued at close on a quarter of a million francs. "Look behind you," said Abdul Ouab. There on the wall hung the portrait. [...] Unable to accept the evidence of his eyes and his hands, Dubois sent for the District Commissioner, the Postmaster and the Hospital Sergeant. Each of them not only saw, but handled the picture. [Then they sent a cable to Dubois' parents in Paris.] At midday he received the following reply: Portrait inexplicably stolen this morning. Police at work. [...] Shortly before sunset Abdul returned and politely

enquired whether the captain had finished with the picture.
The Arab then made a gesture and the portrait vanished. Some
hours later Dubois received a second telegram from his father:
Portrait returned as inexplicably as it vanished.[11]

Is this not amazing? Well, hold on, we will encounter even more fantastic cases!

We have a rather bizarre apport reported in Swami Rama's book *Living with the Himalayan Masters*. Swami Rama tells us of an initiate, a *Haji*, who could apport objects. One salesman told the *Haji* that he would believe him if he could get a heavy Singer sewing machine from his store in Delhi.[12] "[The Haji] said, 'I will get it immediately,' and it appeared. Then the salesman became concerned that it would be missing from the store, and that he might be accused of stealing it. The Haji tried to send it back, but he could not do so. He started weeping and crying, 'I've lost my powers.'" The salesman then had to carry the machine with him; but the missing sewing machine was already reported to the police, and since no one believed the salesman's story, he was accused of theft. Such a cute story! Swami Rama was reprimanded by his master for participating in the theft.

The western equivalents of fakirs and hajis are mediums. There existed many mediums who specialized in apports. The objects that were apported were of various sorts, and, as we will see, in many cases the type of the object itself precluded explanation other than being genuine cases of apports.

Here is a case of a minor apport.[13] During a séance with Mrs Lee Nefertiti, she materialized an apport from the tomb of the Queen of Sheba: a small mosaic bead. Now the story with this apport has a strange continuation. The authors of the referred book visited a psychometry medium, and S. Phillips placed the apport in her hand. Here is what the medium told them: "The chain is modern, but the bead is from an ancient tomb. In fact, I feel that this has come to you as an apport. Could that be possible?" In one fell swoop this one case proves the authenticity of both the phenomena of apports and psychometry. Amazing!

There is a partially successful case of an apport given by Bozzano from which we can infer the mode of production of apports:

In March 1904, in a sitting held in a house of Cavaliere Perreti, in which the medium was an intimate friend of ours, gifted with remarkable psychical mediumship, and with whom "apports" could be obtained "at command," I begged the communication spirit to bring me a small block of pyrites which was lying on my writing-table about two kilometers away. The spirit replied (by the mouth of the entranced medium) that the power was almost exhausted, but all the same he would make the attempt. Soon after the medium sustained the usual spasmodic twitching which signified the arrival of an "apport," but without our hearing the fall of any object on the table, or on the floor. We asked for an explanation from the spirit-operator, who informed us that although he had managed to dis-integrate a portion of the object desired, and had brought it into the room, there was not enough power for him to be able to re-integrate it. He added, "light the light." We did so, and found to our great surprise that the table, the clothes and hair of the sitters, as well as the furniture and carpet of the room, were covered with the thinnest layer of brilliant, impalpable pyrites. When I returned home after the sitting I found the little block of pyrites lying on my writing-table from which a large fragment, about one-third of the whole piece, was missing, this having been scooped out of the block.[14]

In fact, we have a description of the way apports were made, directly from the *guides* (the spirits that control the medium during séances).[15]

Several different guides have explained to us what apports are. They say that every earthly article is composed of atoms, and that they can break up these atoms until the article is entirely disintegrated. The atoms are then transferred en block where they wish them to be, and are materialized again until the article is back in the solid. The process is again a matter of vibrations which enable the de-materialized article to be passed through any solid object.

Now we move to extravagant cases, when the apports were living entities:

In his book Isis in America *Henry Steel Olcott tells us about the medium Mrs. Mary Baker Thayer, who specialized in flower apports during her séances. During one séance Olcott, who was an agriculture expert, counted and identified 84 species of plants. In one case the plant was apparently brought from Scotland, a full-grown heather plant with soil clinging from its roots, and three worms clinging to the soil!*[16]

The Australian medium, Charles Bailey, was more versatile:

The apports included an Indian blanket containing a human scalp [!] and a tomahawk, a block of lead said to be found in Roman strata at Rome and bearing the name of Augustus, a quantity of gravel alleged to have come from Central America and quite unlike anything seen in Australia, two perfect clay tablets covered with cuneiform inscriptions and several thousands of years old, said to have been brought direct from the mounds of Babylon, and finally a bird's nest containing several eggs and the mother bird undoubtedly alive. He was famous for living apports: jungle sparrows, crabs, turtles. Once an eighteen-inches-long shark, at another time a thirty-inch snake appeared mysteriously in the séance room.[17]

Wonderful! During another séance (page 352), "a live shovel-nosed [hammerhead] shark eighteen inches long was brought, [...] also a crab, [...] a quantity of dripping wet seaweed, [...] and live fish." And during yet another séance, "when the lights were turned on, a brown snake, thirty inches [almost one meter] long, was found coiled round the medium's arm, his hand holding it by the neck. It was put on the floor, covered with cloth and in full light immediately disappeared."

In order to pre-empt the trite conjecture that all of the above was the result of trickery by an accomplished sorcerer, we quote from McCarthy's book:

> Sometimes he [Bailey] was stripped and provided with a new suit to wear during the séance. He was then tied up in a large sac drawn tight at the neck with cords, and sealed [...]. His hands were passed through the sides of the bag and tied at the wrists, so that he might take hold of apports of a delicate nature on their arrival. On some special occasions, the sitters searched one another before entering the séance room; male and female sitters retiring separately for the purpose. [...] At the following séance the sitters searched Bailey more particularly; his clothes were removed and his body examined, but, when it was desired to explore the rectum, Bailey took offence, deeming it a gratuitous indignity. The suggestion that live birds [snakes, sharks!] could be stored in the rectum seemed an insult to the intelligence.[18]

An orgy of preparation for a séance! We also learn that a few times this medium was placed "in a cage of mosquito netting, so closely jointed on a wooden frame and secured to the floors with screws that not even a sixpence could be inserted." And "sometimes the door was [locked and] sealed."

Once the apport was in the form of a living lion. It materialized during a séance with the Polish banker, writer, and amateur materialization medium, Franek Kluski. The king of the jungle popping up in a closed room made the sitters uneasy, to say the least:

> *The animal was sometimes rather menacing, beating his tail and striking the furniture. On one occasion the frightened sitters, unable to control the animal, broke up the séance by waking up the medium, "who was deeply entranced."*[19]

One yet another occasion there appeared a kind of a bigfoot or sasquatch:

> *With a mane and a bushy beard, resembling an animal or very primitive man; did not speak, but made hoarse noises, clicked his tongue, and ground his teeth. When called, he approached, allowed his fur to be stroked, touched the hands of the sitters, and scratched them lightly with claws rather than nails. This was an improvement on previous sitting [not well worded], when he was violent and rough. He obeyed the voice of the medium, doing no harm to the sitters.*[20]

Oh boy!

And, lo and behold, these were not the most extraordinary apports during Kluski's séances! That attribute belongs to Kluski himself. Quoting the Italian researcher Ernesto Bozzano via Gwendolyn Kelley Hack's *Modern Psychic Mysteries*:

> *The most extraordinary case related to me by the members of the circle is that of Mr. Kluski having been fetched by apparitions, or disappearing from a sealed and locked séance room. The astonished sitters found him in a rather distant room of the apartment quietly sleeping on a couch. I report the case upon the responsibility of my friends, whom I have no reason to distrust.*[21]

The most often quoted case of a human apport is the instantaneous appearance of the medium, Agnes Guppy. On June 3, 1871, Mrs Guppy was with her companion, Mrs Neyland. At the same time a séance was held with the mediums Herne and Williams, and eight sitters. Here is what happened at the séance:

Mr. Harrison, one of the sitters, asked if they could now bring Mrs. Guppy. The request was half humorous, for Mrs. Guppy was excessively stout. Nevertheless, within three minutes a thud was heard on the table, one of the sitters felt a dress, and a light being struck, Mrs. Guppy was seen standing motionless on the middle of the table, entranced and trembling all over. She had neither hat nor shoes, her right arm was over her eyes, the hand held a pen; and her left hand, hanging by her side, held an account book. [...] When she had recovered they all sat for a séance, during which Mrs. Guppy's boots, a hat, many clothes [...] were brought [...] from her home.[22]

At the other end of this amazing episode, Mrs. Guppy simply vanished:

She [Mrs. Neyland] said that between eight and nine o'clock she was sitting in a room alone with Mrs. Guppy [...], she reading and Mrs. Guppy making up accounts. Once on making a remark there was no reply, and on looking up she was startled to find that Mrs. Guppy was not there. The door was shut, Mrs. Guppy's slippers were on the floor near her chair, and there was a white mist near the ceiling which she, being a medium, recognized as indicating a spiritual manifestation.

Apports of humans happened other times too. Perhaps the most substantiated case is that of a sitter during a séance with Mrs Guppy as the medium. Quoting from Holm's *The Facts of Psychic Science and Philosophy*:

There is also an authentic account [the original source is The Spiritual Magazine, *1874, p.22] of a sitter, a Mr. Henderson [...] being translated from a locked séance room at Mrs. Guppy's residence in London, to the back yard of the house of his friend, Mr. Stokes, a mile and a half distant. It occurred at about ten p. m., on Nov. 2, 1873. There were ten sitters, including Mrs. Guppy, the medium, and while hands were joined in the dark, Mr. Henderson was found to have broken the chain, and, on lighting up, he had vanished from the room, the doors and windows of which were locked. At about the same time he was discovered in the back yard of Mr. Stokes' house, where he was seen by nine members of the household. Although it was a wet night his boots and clothes were practically dry. The fact that ten people observed his disappearance and nine his sudden arrival (all of whose names are appended in the article), makes this account even stronger evidentially than that of the similar phenomenon in which Mrs. Guppy was the principal actor.*[23]

From Batcheldor's scientifically controlled experiment resulting in apports, described in a peer-reviewed publication, through many other controlled experiments resulting in various inanimate and living apports, all the way to human apports, there are degrees of wildness, but the underlying principle is the same.

If taken seriously, which we should, the phenomenon of human apports goes a long way in the direction of establishing what we have set out to establish from the onset: that "we," or rather our bodies, are hallucinations of a sort and thereby, as per the wise Lama Rimpoche, we do not really exist. We will go further in that direction in the remaining three sections of this chapter.

8.3 Bilocation, Doubles

I once saw my wife where she was not! A few moments before the encounter I was deeply immersed in my thoughts while ambling between the aisles of a supermarket. When I came back to my senses within the supermarket scene, I saw my wife by the opposite end of the aisle; she was waving her hand, as if beckoning me to come there. It transpired that she was not there, and she certainly did not wave at me. If that is so, then either I had an encounter with the double of my wife, or I experienced a kind of a timeslip, both *hallucinations* according to the orthodoxy.

At this point we stop for a moment to deliver an appropriate, funny old jingle:

At the turning of the stair,
I saw a man who wasn't there.
He wasn't there again today,
O, how I wish he'd go away!

We defined doubles in the Introduction: they are replicas of the originals — though it is sometimes not clear which is the replica and which is the original. A related notion is that of bilocation; when there are witnesses observing the double, and at the same time, witnesses observing the original. As we mentioned in Chapter 1, there is an obvious overlapping between the two notions, and we will not insist on distinguishing them. From the point of view of orthodoxies, doubles, during bilocation or not, are hallucinations, and so this topic belongs to our main theme.

We have discussed the phenomenon of doubles several times in the previous chapters. For example, in Section 3.1 we presented instances of doubles of dying persons. In this section we will mostly deal with doubles that interact with the witnesses more than just visually.

Simple cases of bilocation/doubles such as my case above have been recorded many times. For example, Reverend S. Baring-Gould saw the bursar (financial secretary) of a college walking in front of him: "I spoke to him. He turned and looked at me, but he passed on without a word."[24] When he encountered the man "again" a little while after, and when he told him that he had just passed him and spoken to him, the bursar turned pale and said he had not left the room at all.

The earliest account of bilocation I am aware of concerns Pythagoras, the Ancient Greek philosopher, mathematician and initiate of a high order. We quote from Porphyry's biography:

> *Almost unanimous is the report that on one and the same day he [Pythagoras] was present at both Metapontum in Italy and the Tauremenium in Sicily, in each place conversing with his friends, though the places are separated by many miles, both at sea and land, demanding a journey of great many days.*[25]

Fast forward a couple of millennia. The original source of the following case is the biography *Life of Loyola* by Francesco Mariani. Ignatius of Loyola, a saint, lived from October 22 1491 to July 31 1556.[26]

> *Leonardo [Clesselis] was a Fleming, and an aged old man, who was the first rector of the college in [Cologne] [...]. He had a most fervent desire again to see the holy father [Ignatius, founder of the Jesuit Order] [...]; he informed him in a letter [begging for permission to walk to Rome.] Ignatius answered that the welfare of others required his stay at Cologne [...], but that perhaps it might please God to content him in some easier way. One day when he [Leonardo] was not asleep, the holy father showed himself to him alive and held a long conversation with him. He then disappeared and left the old man full of the greatest joy [...].*

It has been claimed that sister Maria de Agreda (1602–1665) carried out a "bilocation" from Spain to New Mexico and Texas. The case was investigated by Alonso de Benavides, director of the Franciscan mission in New Mexico at that time (seventeenth century). He confirmed that the local Xumana nation converted to Catholicism en masse after the miraculous arrival of the nun, accompanied by miraculous healing of sick people, and reported on his findings to the King of Spain (1630) and one to the Pope (1634). Moreover, de Benavides interviewed Maria after he returned to Spain, a summary of which was written (and preserved still today) in a letter to the missionaries in New Mexico.[27]

The following is a relatively well-known and well-authenticated case. In September 1774, Alphonsus Liguori, a monk in Arezzo, fell into a five-day long cataleptic trance.[28] When he emerged out of it, he declared that he was with Pope Clement XIV during the Pope's dying hours. Since Rome was an 8-day return trip away, he did not get the information of the Pope's death through ordinary means (not to mention that he did not move during the five days of his trance). This would have been dismissed as an ordinary clairvoyance except that there emerged many witnesses in Rome who actually saw Alphonsus praying by the Pope's bedside. We are told that this case has been well documented and accepted by the Catholic Church as a true bilocation.

A beautiful case of purposeful bilocation was given by the German writer, Jung Stilling. The event occurred between 1750 and 1760. A woman anxious about her seafaring husband, a captain, went to consult a man who is described as being of benevolent, pious character, and suspected of having some occult power of disclosing hidden events.[29]

Having heard her story he [...] went into another room, shutting the door; and there he stayed so long that, moved by curiosity, she looked through an aperture in the door to ascertain what he was about. Seeing him lying motionless, she quickly returned to her place. Some time after, he came out, and told the woman that her husband was at that time in London, in a certain coffee-house which he named, and that he would soon return. [...] When her husband did return they found, on comparing notes, that every thing she had been told was exactly true. [...] When she took her husband to see the alleged seer, he started back in surprise, and afterward confessed to his wife that, on a certain day (the same on which she had consulted the person in question,) he was in a coffee-house in London (the same that had been named to her) and that this very man had there accosted him, and had told him that his wife was in great anxiety about him; that then the sea-captain had replied informing the stranger why his return was delayed and why he had not written, whereupon the man turned away, and he lost sight of him in the crowed.

Of course, this case is weak from the "scientific point of view." We gave it anyway.

A cute story is the following case of bilocation by Padre Pio:

Once, just before going on the air, the announcer's [an Italian radio announcer] head ached violently that he was temporarily paralyzed. A few seconds later Padre Pio came into the studio, put his hand on the man's forehead, and the headache vanished. Astonished as the announcer was, he later decided that it must have been a hallucination. He went to see the priest to tell him what had happened, but before he could open his mouth, Pio put his hand on his visitor's forehead and said, smiling, "Oh, oh, these hallucinations."[30]

The most amazing cases of bilocation are those where talented people produce functional doubles, directing them to chosen places. The case of Peter Lärdal, a Lap shaman, was given earlier (Section 5.2).

Here is another anecdote involving purposeful bilocation:

In 1923 Mrs Jensen concentrated on her husband who was away in an unknown to her place, on a business trip. She "found him" walking down an alley, and followed him until he entered his room and undressed to go to bed. At the same time, Mr. Jensen did precisely what his wife eventually described (matching the scene of the place she has never visited). When he returned to his hotel he "suddenly saw the figure of his wife standing beside his bed."[31]

The next two cases are rather intriguingly similar.[32]

W. T. Stead describes in Borderland *(Vol. III, p.26), the case of his friend, Mrs. A., whose double, while she herself was in bed, very ill, attended an evening service at Stead's church (on October 13th, 1895), where it was seen by many and recognized by Stead and four others. Mrs. A.'s double entered the church during the first hymn, walked up the aisle and entered a vacant pew next the choir. A hymn book was handed to her by a lady, which she took, but she did not appear to sing, and sat perfectly still throughout the service. [...] She sat still until the singing of the last hymn, when she stood up, holding her hymn book; at the last verse she laid the book down, walked quickly down the aisle, opened the door herself and passed out.*

Stead obtained statements from those who had seen her at home, and those who had seen and recognized her double in the church. Mrs A. told Stead — and what she said was corroborated by the doctor and others in the house — that she slept between 7 and 8:30, that she had not thought of the church or wished to be there and had no consciousness of having been there.

The second case happened in the House of Commons in England, and it was described in several newspapers of the time (1905), including *The Daily News* of May 17, 1905. Sir Carne Rasch was sick and could not make it for a very important vote in the commons. Nevertheless, he was seen by at least two people. Here is the account of Sir Gilbert:

> *My gaze fell upon Sir Carne Rasch seated near his usual place. As I knew he had been ill, I waved to him in a friendly way, and said: "I hope you are better." But he gave no sign of recognition, which greatly astonished me. His face was very pale. [...] For a moment I wondered what I had better do; when I turned toward him, he had disappeared.*[33]

Meanwhile, of course, the ill Sir was lying bedridden in his home.

A very amusing and consequential incident was recounted to the writer Robert Dale Owen, second hand by Captain Clarke, who in turn was told of it by Mr Bruce who sailed with Clarke for seventeen months, between 1836 and 1837. The event happened eight years before Bruce told the captain of it, and twenty years before the captain retold the story to Owen. In short, Bruce was the first mate of a barge trading between England and Canada, when one day, while he was computing the position of the ship, he saw a man who was not anyone from the crew of the ship. The man was writing something on a slate. The captain was called, and the slate indeed contained a short sentence: "Steer to the northwest."[34] The handwriting did not match any of the handwriting of the people on board. The captain followed the request written on the slate, steered to northwest and, shortly after, they found a passenger ship from Quebec, entangled in ice for two weeks. The grateful passengers were all saved, and among them Bruce recognized the apparition he had seen in the captain's cabin. The man was not aware of his "double" visiting the ship. However, as the captain of the wrecked ship confirmed, at that time he was sleeping, and upon waking up informed the captain of his certainty that they would be saved. The clairvoyant passenger then described the ship that would eventually indeed save them. Cute!

The following bizarre story was told by Camille Flammarion, whom we have quoted several times throughout this book. A Canadian visited Flammarion and told him that three years ago he had abandoned his wife and children in order to fulfill his quest of understanding and verifying the reality of certain phenomena through his own personal experience.[35] He had been assured that such facilities would be acquired if he (1) abstained from eating meat and fish; (2) ate only vegetables that he himself cultivated, gathered and cleaned; (3) drank only water; (4) preserved absolute chastity; and (5) planned his days according to certain rules given to him (presumably he was asked to meditate daily). Well, he succeeded! The morning he visited Flammarion he saw his own double stretched on the sofa. His double then went to the window of his apartment on the fourth floor and was about to jump down to the boulevard. That was too much for the Canadian, and he decided to give up the quest and go back to his family.

The most amazing case of bilocation is given in Baird Spalding's seminal book *Life and Teaching of the Masters of the Far East*, Volume 1. "Jast," one of Spalding's guides, led Spalding (and a few other people) to the forest nearby:

> We stood staring as though transfixed for we saw that the figure lying on the ground was Jast. Suddenly, as Jast walked toward it, the figure became animated and rose to a standing position. As the figure and Jast stood face to face for an instant, there was no mistaking the identity — it was Jast. Then, instantly,

the Jast we had known had disappeared and there was but one
figure standing before us. [...] It was very evident that Jast's
body had been lying where we found it for a considerable time.
The hair had grown long and bushy and in it were the nests of
a little bird peculiar to the country.[36]

We digress to note a momentous thesis regarding the secret of the "philosopher's stone." Recall that the philosopher's stone is a metaphor for achieving an everlasting and healthy life. We now know what needs to be done: produce, by yourself or with some generous help of the invisible powers, an upgraded replica of yourself, perhaps using a younger facsimile, free of maladies and other imperfections, and transfer all of your knowledge, memories and personality to this double. Then dispose of the original, preferably by vanishing it altogether. That's it!

The first step — the creation of a double — is immense! It requires true knowledge of self, combined with generous support of the invisible entities who control our destiny. To that goal, one does not merely request assistance from the entities in the higher spheres of existence. It is necessary to find them, including The Supreme Consciousness, within one's deep self. Lust for life and fear of death are counterproductive in this context. In fact, the opposite may be needed: abandonment of body-linked individuality and the total resignation of personal will to the will of the hierarchy of higher selves, all the way to the Supreme Consciousness or God.

We recounted just a few out of the substantial library of cases of doubles/bilocation; the phenomena occurred just too many times to be easily dismissible. The double is, according to our definition, a prime example of a hallucination. In the extreme cases, when there is hardly a difference between the double and the original, it follows that we are also hallucinations! We will substantially strengthen this thesis in the next section.

8.4 Materializations and Dematerializations

The material in this section is arguably the most outlandish compared to what we have encountered so far — if that comparison could be made at all given the surrealness of virtually everything we have discussed. We present cases of materialization and dematerialization of human forms, paying special attention to cases of materialization of fully functional and apparently solid human entities. Such cases were abundant products of séances held during the second half of the nineteenth century and the first quarter of the twentieth century.

The cases of full materialization used to be so ubiquitous, that the British writer, H. Dennis Bradley, frequently expressed his boredom with mediums specializing in matter-related supernatural phenomena. He much preferred chatting with *spirits*. He wrote:

My considered conclusion, reached after a prolonged series of experiments and investigations, [...] is that two facts of colossal importance to the human race have been established beyond cavil. First, that there is a survival of the spirit after bodily death, and second, that it is possible for living people to enter into direct communication with those who have passed away.[37]

Direct communication, or *direct voice* phenomena happen when there is an articulate voice emanating from a point in the séance room that is some distance away from the sitters and the medium. Bradley chatted many times with (the voice of) his dead sister during séances with the American medium, George Valiantine.

Bradley did not mince his words: "I know that there is no death, but there are degrees of life, and it is the spirits who are completely alive and we who are comatose."[38]

The human forms that appeared during séances were at times ostensibly made of *ectoplasm,* a slimy whitish substance emanating from the mediums, primarily from the mouth but also from ears, nose, eyes, breasts, and lower orifices, as well as from other places where there was no aperture (the top of the head, and from the fingertips). The existence of this substance emanating from mediums is beyond any reasonable doubt: there are numerous photographs and videos confirming the phenomenon. Moreover, the mediums were at times unreasonably stringently controlled, including thorough checks of their bodies before the beginning of the séances. For example, before the sitting of 29 December 1910, the medium Eva C. (Marthe Beraud), was more than thoroughly checked. Said Baron and Doctor Albert Von Schrenck Notzing: "Mme Bisson, in my presence, introduced her finger into the medium's vagina. She was also explored by Professor B. and the author through the garment, but with negative result."[39] As if a little fairy (see the next example) can be hidden in a vagina. Eva C. was thus unceremoniously demoted to less than a guinea pig, all in the name of science.

During séances with Eva C., there appeared numerous materialized forms of human parts. At times these forms were two-dimensional, resembling cardboard cut-outs. Sometimes they were materialized whole human forms:

Mme. Bisson describes [in Review Métapsychique, *1921, p.364] how on May 25th, 1921, at 4:30 p.m., [during a séance with Eva C.] a beautiful little naked female, only 8 inches [20 cm] high, fashioned itself gradually from a mass of ectoplasm [material emanating from Eva C.] on the medium's hand placed outside the sack [in which Eva C. was put; there were holes in the sac for hands] in full view. It had long hair and its skin was brilliantly white in the daylight which flooded the room. It vanished, but reappeared immediately with the hair differently arranged. It again disappeared but returned and now was only 6 inches (15 cm) high. Its anatomy seemed to be perfect, for it went through various gymnastic exercises, and finally stood on Mme. Bisson's hand. There were six sitters at this particular séance, and all had ample opportunity of observing the little figure.*[40]

Given that, as we have seen earlier, miniature fairy-like human forms that are replicas of dying persons have been observed many times, it is no surprise that similar phenomena happened during séances with mediums.

The materialized human forms observed numerous times during mediumistic séances represented deceased people. Most of the time these "dead" people lived mundanely long ago and were thereby not traceable through written records. However, at times there manifested, in fully materialized forms, deceased persons known to some of the witnesses. Consider, for example, the case of "Rosalie," born out of a personal tragedy. It is rather beautiful and touching, so we devote a few paragraphs to it. It comes from Harry Price's book *Fifty Years of Psychical Research*.

A French lady, called Madame Z. by Price, married an Englishman who afterwards was killed in the First World War. They had a daughter, Rosalie. Rosalie died at the age of 6. Quoting from page 134 in Price's book:

In the spring of 1925 [four years after Rosalie passed away] Madame Z. was awakened during the night by the sound of her dead girl's voice crying "mother." This occurred so frequently that Madame Z. got into the habit of lying awake at night, waiting for the "voice." Gradually, she thought she could see (in the dark) the dim outline of "Rosalie" and hear her footsteps in the room. Finally, the mother declared, one night she put her arm out of bed and her hand was clasped by that of her little girl.[41]

Madame Z. lived mostly a solitary life. Her main company was a middle-class couple from the neighborhood, Mr X and Mrs X. In 1928, the couple proposed to Madame Z. that they organize a sitting for Rosalie. They did not get any response during the first six months. In the spring of 1929, Rosalie materialized for the first time, and then she started appearing regularly. Mrs X, who had read a published version of Price's broadcast talk, approached Price eight years after, in December 1937. Quoting from Price's book *Fifty Years of Psychical Research*: "She told me that she had noted that I could 'guarantee a ghost' in a particular haunted house which I mentioned in my broadcast; she, too, could 'guarantee a ghost,' but one of a much more objective nature than any I had experienced."[42] So, under some conditions (say, not to ask questions or disturb the apparition in any way during the séance without permission; Madame Z. was afraid that Rosalie would not show again if disturbed) Price agreed to participate in a séance. He was given free hand to check and do whatever he wanted before and after the séance. And so, he did. He diligently double checked the entire house, searched the people involved in the sitting (Madame Z, the Xs, and their perspective son in law); he sprinkled starch powder around the doors, windows and the chimney, he locked and sealed the doors and the windows, and he removed some of the furniture from the séance room. And the séance started. They waited for an hour or so, sitting quietly, with occasional whispers by Madame Z., calling Rosalie, while she and Mrs X were quietly sobbing. Then Rosalie appeared in the dark room.[43]

> *The next sound I heard was a sort of shuffling of feet on my left [where Madame Z. was sitting; the sitting arrangement was also decided by Price] at the same moment as something slightly touched the back of my left hand, which was resting on my knee. It felt soft and a little warm. [...] Madame Z. continued to whisper at the "child," and her sobbing ceased somewhat.*

[He was given permission to touch the materialization.] I stretched out my left arm and, to my amazement, it came in contact with apparently, the nude figure of a little girl, aged about six years. I slowly passed my hand across her chest up to her chin and cheeks, Her flesh felt warm, though not so warm as one would expect to find normal human flesh. I laid the back of my left hand on her right cheek; it felt soft and warm and I could distinctly hear her breathing. I then placed my hand on her chest again and could feel the respiratory movements. My hand travelled to her tights, back and buttocks, then traversed her legs and feet. They were normal the limbs of a normal six-year-old. [...] I could feel her hair, long and soft, falling over her shoulders. [...] A supreme scientific interest, with a feeling of absolute incredulity would best describe my reaction. I had not bargained for anything so wonderful (or so clever!) as this. [He is more concerned about covering his back than opening his "scientific mind."] [...] With my right hand, I lifted "Rosalie's" right arm and felt her pulse. It appeared to be too quick and I estimated a rate of 90 to the minute. I put my ear to her chest and could distinctly hear her heart beating. [...] At this juncture I asked my hostess if Madame Z. would allow me to use the luminous plaque. After a little discussion it was agreed that both Mrs. X and I should shine our plaques on "Rosalie," the stipulation being that we should begin at the feet of the form, and then later illuminate the upper part of the child. I picked up my plaque and in turning it over a soft, fluorescent flow flooded the feet of "Rosalie." They were normal feet of a normal child. [...] I could see the soft texture of the flesh, which appeared to be without a blemish. As our plaques traveled upwards the face of the form was revealed and we beheld a beautiful child who would have graced any nursery in the land. Her features were classical and she looked older than her alleged years. [...] Her eyes (they appeared to be dark blue) were bright with an intelligent gleam in them. [Price

then requested to put some questions to "Rosalie," and he was given permission with the remark that that it was unlikely he would get answers. He asked the following: "Where do you live, Rosalie?" "What do you do there?" "Do you play with other children?" "Have you any toys there?" "Are there any animal pets?" and got no answer.] "Rosalie" simply stared and did not seem to understand what I was saying. I asked her a final question: "Rosalie, do you love your mummy?" I saw the expression on her face change and her eyes light up. "Yes," she lisped. "Rosalie" had barely uttered this single word when Madame Z. gave a cry and clasped her "daughter" to her breast. Mrs. X. [...] asked for completely silence — rather difficult as all the women in the circle were crying. I must admit that I was rather affected myself — it was a touching and pathetic scene.

I would call it poignantly touching! I would classify this as materialization generated by Madame Z. after many years of deeply emotional "labor." Note that, unlike the manifestations of deceased children mentioned in the previous section, "Rosalie" did not age. Madame Z. longed for her 6-year-old daughter, not a daughter who had aged out of her sight.

More often than not, the apparition of dead children seen years after their dying show ageing corresponding to passed time. Here is an example. Florence Marryat went to a séance, completely unknown, and the attendant simply showed her the way to the séance room, without a word being said. There were about 30 to 40 people attending the séance; the medium was a certain Mrs M. A. Williams, and materialization was her specialty. Here are two segments of the séance.[44]

I don't think it could have been more than a minute or two before we heard a voice whispering, "Father," and three girls, dressed in white clinging garments, appeared at the opening in the curtain. An old man with white hair left his seat and walked up in the cabinet, when they all three came out at once and hung about his neck and kissed him, and whispered to him. I almost forgot where I was. They looked so perfectly human, so joyous and girl-like, somewhere between seventeen and twenty, and they all spoke at once, so like what girls on earth would do, that it was most mystifying. The old man came back to his seat, wiping his eyes. "Are those your daughters, sir?" asked one of the sitters. "Yes! My three girls," he replied, "I lost them all between [seven and] ten years old, but you see I've got them back again here." [...]

Once the conductor spoke to me. "I am not aware of your name," he said [...], "but a spirit here wishes you would come up to the cabinet." I advanced, expecting to see some friend, and there stood a Catholic priest with his hand extended in blessing. I knelt down and he gave me the usual benediction and then closed the curtains. "Did you know the spirit?" the conductor asked me. I shook my head; and he continued, "He was Father Hayes, a well-known priest in this city. I suppose you are Catholic?" I told him yes and went back to my seat. The conductor addressed me again. "I think Father Hayes must have come to pave the way for some of your friends," he said. "Here is a spirit who says she has come for a lady named Florence, who just crossed the sea. Do you answer to the description?" I was about to say "Yes," when the curtain parted again and my daughter, "Florence" ran across the room and fell into my arms. [Her daughter then kissed her, they spoke a little, and then she told her mother that another spirit was there for her.] She was going back to the cabinet when the conductor stopped her. "You must not return this way, please. Any other you like," and she immediately made a kind of court curtsey [formal greeting made by bending the knees with one foot in front of the other] and went down through the carpet. [A moment later] she came up again a few feet from me, head first, and smiling as if she had discovered a new game. She was allowed to enter the cabinet this time, but a moment afterwards she popped her head out again, and said, "Here's your friend, Mother!" and by her side was standing William Eglinton's control, Joey, clad in his white suit, with a white cap drawn over his head. [She then chatted with both spirits, and Joey told her that he must go back to Willy (Eglinton).] I really didn't care if he stayed long or not. I seemed to have procured the last proof I needed of the truth of the doctrine I had held so long, that there is no such thing as Death, as we understand it in this world.

The second half of the nineteenth century were the times when extremely talented mediums could have been readily found in the New York newspapers adds. One can say that it was the era of spiritual freedom in the West.

Case number 100 from the book *Casebook for Survival* by Alex Baird, possibly overlaps with the previous case given here. Florence Marryat lost a child, also called Florence, when the baby was ten days old; the child had deformities on her face, and her gullet was missing. Ten years later, Florence anonymously attended a séance with a certain Mrs Holmes, who reported the presence of a girl, her face covered by ectoplasmic substance. The child claimed a relationship with Florence, which Florence denied. This scenario repeated several times with different mediums, always anonymously. Years later Florence visited a trance medium.[45] "In the middle of the sitting the control was suddenly changed and Florence was startled by hearing: 'Mother, Mother. I am Florence. I want to feel I have mother. I'm so lonely...' [Florence the mother] 'but I always think of you my dear, dead baby.' [Florence the baby] 'but I'm not the baby now...'" About 17 years after the birth of her baby, Florence attended a sitting with the medium Florence Cook: "... fully materialized girl crossed the room and sat on her mother's lap. [She told her mother to turn up the lights.] They all saw distinctly that particular defect on the lip with which she was born [...]. She also opens her mouth so I might see if she has no gullet. [...] [Florence or younger] "Don't fancy I really look like this now; the blemish left me long ago, but I put it on tonight to make you certain." The materialization lasted for 20 minutes! Kids growing up in the postmortem or spirit realm?[46]

Here is a more recent example: "Three years after Joseph's death, Richard [his father] was by himself in church on a Saturday evening after attending mass. 'I saw two boys [his two deceased sons], but they were not the ages they were when they died. [...] They were kneeling beside me.' Nothing was said."[47]

According to the psychologist, Michael Jackson, [...] who worked at the Alister Hardy Research Centre in Oxford, Richard's experience of seeing his sons at the age they would be if they had lived was not unusual.[48] "He [Jackson] has observed in a number of accounts that when people lose a child and then see the child's ghost several year later, he or she will often have aged by the correct number of years." This is a re-occurring claim regarding manifestations of children during mediumistic séances. For example, an entity — or a *spirit* — that manifested through direct voice during a séance in 1925 with the medium Valiantine, stated emphatically that a child "was growing up in the spirit world."[49] This is compatible with our theory (Chapter 6) that the Deep Reality contains all alternative outcomes of all events, and it supports my own experience described in the last section of the book (Epilogue).

Here is a description of the process of materialization of a "spirit" during a séance with the medium William Eglinton (1880), by Eglinton's biographer, John S. Farmer:

Then, standing in full view, by a quick movement of his fingers, he [Eglinton] gently drew forth, apparently from under his morning coat, a dingy white-looking substance. He drew it from him at right angles and allowed it to fall down to his left side. As it reached the ground it increased in volume and covered his left leg from knee downwards. The mass of white material on the ground increased in bulk and commenced to pulsate, move up and down and sway from side to side. Its height increased and shortly afterwards it quickly grew into a form of full stature, completely enveloped in the white material. The upper part of this the medium then drew back and displayed the bearded face of a full-length materialized spirit, considerably taller than himself. [...] The enveloping white material was now seen to be a flowing robe, fastened around the waist with a girdle. After a few minutes the medium, still in trance, drew forth more of the white material and stretched it out to the spirit which eagerly grasped it. Finally, the medium became weak, staggered and was supported by the nearest sitter, whereupon the spirit approached and dragged him into the cabinet.[50]

We have a similar account; the original source is A. R. Wallace's book *My Life*:

It was a bright summer afternoon, and everything happened in full light of the day. After a little conversation, Monck [the medium] appeared to go into a trance; then stood up a few feet in front of us, and after a little while pointed to his side, saying, "Look." We saw there a faint white patch on his coat on the left side. This grew brighter, then seemed to flicker, and extend both upwards and downwards, till very gradually it formed a cloudy pillar extending from his shoulders to his feet and close to his body. Then he shifted himself a little sideways, the cloudy figure standing still, but appearing joined to him by a cloudy band at the height of which it had first begun to form. Then, after a few minutes more, Monck again said "Look," and passed his hand through the connecting band, severing it. He and the figure then moved from each other till they were about five or six feet apart. The figure had now assumed the appearance of a thickly-draped female form, with arms and hands just visible. Monck looked towards it and again said to us "Look," and then clapped his hands. On which the figure put out her hands, clapping them as he had done, and we all distinctly heard her clap following his, but fainter. The figure then moved slowly back to him, grew fainter and shorter, and was apparently absorbed into his body as it had grown out of it.[51]

Spirit entities behave in many ways like apparitions, often vanishing by sinking into the floor, or upwards through the ceiling. At times the manifestations were rather spectacular. For example, here is an account by Mrs Crookes, the wife of Walter Crookes, of a séance with D.D. Home as the medium, on March 9 1893:

The accordion began to play, and the figure [a cloudy appearance which soon condensed into a distinct human form] advanced towards me till it almost touched me, playing continuously. It was semi-transparent, and I felt an intense cold, getting stronger as it got nearer, and as it was giving me the accordion I could not help screaming. The figure immediately seemed to sink into the floor to the waist, leaving only the head and shoulders visible, still playing the accordion, which was then about a foot off the floor. Mr. Home and my husband came to me at once, and I have no clear recollection of what then occurred, except that the accordion did not cease playing immediately. Mr. Serjeant Cox was rather angry at my want of nerve, and exclaimed: "Mrs. Crookes, you have spoilt the finest manifestation we have ever had."[52]

One of the cases most often quoted in books is that of the materialized spirit named Katie King, manifesting during séances with the medium, Florence Cook. The prime investigator of this case was Sir William Crookes, an eminent physicist of his time. Crookes initially expected to discover fraud and trickery, and his announcement that he would scrutinize the phenomena of spiritualism was received with jubilation from the establishment. However, "foregone conclusions have never met with more bitter disappointment. [...] Crookes' report was submitted to the Royal Society on June 15, 1871, but his communications, as they did not demonstrate the fallacy of the alleged marvel of Spiritualism, were refused [...]."[53]

Katie King materialized while Florence Cook was entranced, isolated in a cabinet. She was completely solid, walked about the room, chatted with the sitters, and, in general, behaved like a normal human being. She even held in her hands an infant child of Crookes, exhibiting motherly affection. Since Katie King physically strongly resembled Florence Cook, the obvious inference was that the former was nothing more than a theatrical role played by the latter. However, it was not that simple. Crookes was a thorough investigator; he once attached a galvanometer to the medium, registering the slightest movement by her. Nevertheless, Katie King appeared, and the meter needle never moved. Katie King was substantially taller that Florence Cook, she did not have pierced ears while the medium had. Significantly, several times both Cook and King were seen at the same time; indeed, Crookes witnessed the touching farewell meeting between Katie and Miss Cook during the last séance, when the two were talking affectionately, Miss Cook shedding many tears.

Katie King could do things that no "normal" human could replicate. For example, "When the sitters complained that her [Katie King's] face was like the medium's [Florence Cook], she [the materialized spirit Katie King] said that she could not help it, but showed her power by changing its colour from white to chocolate colour, and then to a jet–black which shone like patent leather. She did this several times."[54]

Instant changes in the appearance of materialized "spirits" happened many times during séances. For example, "Mr. Clemens describes in *Light* for 1894 (p.77) a séance with Mrs. Aspinall as medium, in which a friend of his, an officer, materialized. On this occasion the form was in evening dress, and on being asked why he was not in his uniform as on a previous occasion, he at once converted his evening dress into military uniform by merely making a few passes over his body."[55]

Back to Katie King — her materialization was a fully formed woman: "I [Florence Marryat] asked her [the materialized spirit Katie King] if [...] blood ran through her body, and she had a heart and lungs. Her answer was, 'I have everything that [the medium] Florrie [Florence Cook] has.' On that occasion also she called me after her into the back room, and, dropping her white garment, stood perfectly naked before me. 'Now,' she said, 'You can see that I am a woman.' Which indeed she was, and a most beautifully made woman too."[56]

Nevertheless, at times she acted as a typical hallucination, vanishing or melting away at various moments. Quoting from Florence Marryat's *There is no Death*: "Then three gas-burners were turned on to their full extent [...]. The effect upon Katie King was marvellous. She looked like herself for the space of a second only, then she began gradually to melt away. [...] First, the features became blurred and indistinct; they seemed to run into each other. The eyes sank in the sockets, the nose disappeared, the frontal bine fell in. Next the limbs appeared to give way under her, and she sank lower and lower on the carpet like a crumbling edifice. At last there was nothing but her head above the ground, then a heap of white drapery only, which disappeared with a whisk [flick] ..."[57]

The phenomenon of materialized spirits during séances with mediums was very common, and at times large numbers of different spirit entities manifested. For example, during a single séance with mediums William and Horatio Eddy there appeared as many as twenty materialized spirits of all ages. Colonel Olcott states[58] that he saw several hundred different spirits during his stay with the Eddys! Similarly, from twelve to twenty different spirits materialized *each evening* during twice weekly séances with the powerful medium George Spriggs between 1877 and 1879.

Materializations of fully functional replicas of ostensibly dead humans, or *spirits*, have also occurred outside the scope of séances with mediums. Usually there was a good reason for such manifestations, as was the case with "Rosalie" and Madame Z. In some cases, the impetus seems to have come from entities from a higher domain of existence, sometimes referred to as higher planes. For example, there is the case of the notable psychiatrist, an inaugurator of near-death studies, and humanist Elisabeth Kübler-Ross. During the time of a crisis in her life, when she seriously considered giving up her work, she was persuaded to stay on course by no less than a patient of hers who passed over almost a year earlier! The patient, Mrs Schwartz, approached Kübler-Ross after what was to be her last seminar, and encouraged her to continue with her work with dying patients. Kübler-Ross perceived Mrs Schwartz as somewhat transparent. Since she could not believe herself what she was seeing, she asked Mrs Schwartz to write a note and sign it, and the signature and the handwriting was subsequently verified to have been that of the living Mrs Schwartz.

By the way, many years later during an interview, Kübler-Ross was asked if she believed in life after death, she replied, "I don't believe; I know. I know that life does not stop at death."

In several of the above cases, the materialized forms were fleeting and ephemeral, and in general hallucination-like. We will now exhibit the most far-fetched cases of materialization, indicating that these attributes apply to us too!

There are many cases of trance-mediums, who were usually isolated in their cabinets, altogether disappearing. Typically, this happened while the mediums were strictly controlled, sometimes by electronic means — recall the galvanometer attached to Florence Cook, mentioned earlier in this section. It has been noted that the medium, Mrs Compton, *always* entirely disappeared during a materialization, leaving no trace, though she was carefully fastened to preclude free movement. The left arm of Indridi Indridason, the Islandic medium whose mediumship was brief and lasted only four years (from 1905 to 1909), disappeared during three séances. Nielsson, in *Wonderful Boy Medium in Iceland*, 1919, states that seven persons observed the phenomenon the third evening that Indridason's left arm disappeared.[59]

> They shone light all around Indridason while the empty sleeve hung down. They were permitted to touch and feel him all round the shoulder. [...] Indridason stood on the floor and Kvaran felt from Indridason's "shoulder down along his side and back and the same on his front." He also felt "all around him and high and low over the trunk." [...] All the seven witnesses signed a document at the end of this séance, stating that they had not been able to feel or find Indridason's arm and were prepared to certify this under oath.

By the way, we are told that "Indridason was an uneducated son of a farmer and knew no foreign language except for a few words in Danish. Nielson [...] writes that the young country boy had never seen a juggler, and conjuring tricks were at the time quite unknown to Iceland."[60]

We end our brief review of cases of partial dematerialization with four fascinating instances. The first is that of the medium Mme D'Esperance (Elizabeth Hope); it happened on December 11, 1895. It is described in Aksakof's booklet *A Case of Partial*

Dematerialization, 1898, that included the testimonies of fifteen witnesses.[61]

> *[The] lower part of the medium's body, from the waist downward disappeared. Her skirt was lying flat on the chair for about 15 minutes and the medium's trunk was apparently suspended in the air above the seat. The light was sufficient to see by and Mme. d'Esperance permitted five persons to verify the phenomenon by passing their hands below the trunk. [...] Mme. D'Esperance's account of her feelings is especially interesting. She said: "I relaxed my muscles and let my hands fall upon my lap and then I found that, instead of resting against my knees they rested against the chair in which I was sitting. This discovery disturbed me greatly and I wondered if I was dreaming. I parted my skirt carefully, all over, trying to locate my limbs and the lower half of my body, but I found that although the upper part of it [...] was in its natural state, all the lower part had entirely disappeared. [...] Nevertheless, I felt just as usual — better than usual, in fact." [...] Fifteen minutes later her skirt filled out and her lower limbs appeared in full view of the sitters.*

The Brazilian medium, Mirabelli, whom we mentioned in Section 7.6, disappeared in broad daylight while tied down in an armchair and watched closely by a group of doctors.[62]

Quoting from Fodor's encyclopedia: "[In around 1875, the writer] William Oxley once observed a total dematerialization of the [English] medium [Mr. Firman] while the phantom was outside the cabinet."[63]

The fourth fascinating dematerialization case is described by Florence Marryat in her book *The Spirit World*, 1894. She was led by the materialized spirt "Florence" behind the curtains to see [the medium] Miss Showers. She writes:

The first sight of her terrified me. She appeared to be shrunk
to half her usual size and the dress hung loosely on her figure.
Her arms had disappeared, but putting my hands up the dress
sleeves I found them diminished to the size of those of a little
child — the fingers reaching only to where elbows had been.
The same miracle had happened to her feet, which only occupied
the half of her boots. She looked in fact like a mummy of a girl
of four or six years [...].[64]

The source of the flow of matter into the materialized spirits does not end with the medium.[65] "Dr Crawford [found] that the sitters also contribute to the ectoplasmic flow. He discovered it by measuring the variation in weight during the séance of both the medium and the sitters."

The weights of the mediums had been measured many times, and it had been conclusively shown that it varies considerably during materialization séances. So did the weights of the materialized spirits: "[During a séance with Spriggs] different spirits were weighted. One weighted 139.5 pounds, another 33.5 pounds. The spirit 'Lily' altered her weight at will; taken successively it was 53.25, 45, 34.5, and 33.5 pounds. The medium weighted 148.5 pounds."[66] So much for the accusation that mediums impersonate materialized spirits!

It is interesting that the variable weight of the mediums was sometime considered as an incriminating factor. The French psychologist and researcher into paranormal phenomena, Charles Richet, dismissed the German medium, Ana Rothe, as a fraud because she weighed 1 kilogram less after a test during which she *materialized* flowers, which was exactly the weight of the flowers. From this, Richet concluded that she had smuggled the flowers. Apparently, the possibility that the extra kilogram was used to bring about the flowers did not occur to Richet. Ana Rothe was arrested during another séance, because, as it was claimed, the product of materialization was nothing more than natural flowers which were hidden beneath her skirts! Despite the testimony of many witnesses attesting to the genuineness of her mediumistic talent, she was sentenced to eighteen years of prison.

We end this section by noticing that not all materializations or other phenomena contradicting the presumed solidity of matter are brought about by mediums. Recall that a large percent of widows and widowers experience fully materialized hallucinations of their spouses. We also have the truly fascinating case of Dorothy Eady, or Omm Sety, who made love with the fully materialized body of the pharaoh Seti, whose life happened more than 3000 years ago.[67] And we mention the vast lore related to the world of fairies, relegated to fiction by modern society. It contains many tales and anecdotes according to which the solidity of matter is illusory. Here is a simple example, coming from Janet Bord's *Fairies: Real Encounters With Little People*, page 192.

> *Reverend Dr. A. T. P. Byles and his wife found a hole in the path in the churchyard. It was about a yard wide, and when the vicar threw a stone down it, he heard it hit stonework. They hurried off to fetch planks to cover the hole, but when they returned there was no hole to be found; the path was normal.*

This happened in the 1940s, well past the period ending with the nineteenth century, when such occurrences were relatively frequent.

8.5 Reality: Concluding Remarks

Do the cases we have covered in this chapter fall within the category of "impossible miracles"? Hardly so! Let's resort briefly and once again to the String Theory of relatively modern physics.

Recall that according to this theory the most elementary units of matter and energy are strings that vibrate. Different frequencies of vibration give rise to different manifestations of matter or energy, so that the term "spectrum of matter" makes sense.

The associated mathematical differential equations precisely modeling the theory are rather complicated even when there is only one oscillating string. It is possible that the actual elementary units of matter consist of many strings that oscillate at different rates at the same time. This is also implied by the clairvoyant version of String Theory. In the year 1870, the American physician and scientist Edwin D. Babbitt (1828–1905), one of the discoverers of Chromotherapy (color therapy), "commenced cultivating, in a dark room and with closed eyes, [his] interior vision." After a few months he managed to see (clairvoyantly) the structure of the most elementary subatomic particles of matter. Accordingly, they are multiple strings of energy linked in knotted spirals; each of these energy spirals is made of smaller spirals winding around them; each of the smaller spirals is made of even smaller spirals winding around them; and this is repeated the total of seven times (as illustrated).[68]

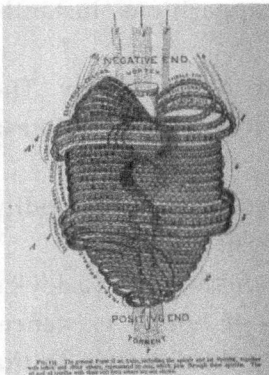

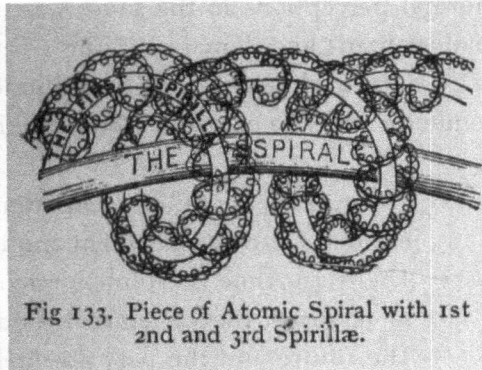

Fig 133. Piece of Atomic Spiral with 1st 2nd and 3rd Spirillæ.

Subatomic strings according to Babbitt.

In 1895, the clairvoyant Charles Leadbeater was requested to try to see, also clairvoyantly, the structure of hydrogen. He (allegedly?) succeeded and described it as a certain configuration of linked spirals that were virtually identical to Babbitt's.[69] Should this be dismissed as occultist junk? Perhaps not. The same experiment in clairvoyance produced a description of some atoms that was subsequently verified. Leadbeater saw that a hydrogen atom could have one, two, or three "particles" in its nucleus, and still be hydrogen. Neutrons (the "particles") and the isotopes of hydrogen (deuterium with two neutrons, and tritium with three neutrons) were discovered some 35 years later (early 1930s).

Back to our "miracles." To make a part of the human body imperceivable to humans, the *only* parameter that needs to be changed is the rate of oscillation of the associated strings. The technology of this change is based on the dominance of mind over matter, and it is mainly in the possession of superior entities out of this world. Literally! But the concept is accessible to us. The idea that parts of human bodies, or all of it, can be made invisible or imperceivable, becomes less than extravagant. And with it, there goes hand in hand the thesis that we are hallucinations of a sort; we enter and exit this world of our normal perceptions in the same way as apparitions, instantly materializing and dematerializing.

We have arrived at the final roundup of our theory. The evidence we have presented led us to the following postulates:

- There is no death; our consciousness transcends bodily existence and is permanent and indestructible.
- There is no time, and our experience thereof is just a device to navigate the ephemeral modes of bodily existence. The future and the past are merely thin streams in the magnificently colossal ocean of true time, the spacious

present. The Universe is a huge constant containing all possible time streams, only some of which are being actualized.

• We don't really exist. Our bodily existence is as substantial, or insubstantial, as are the apparent hallucinations. We hallucinate ourselves into such existence.

We revisit the definition of hallucination from Section 1.1: accordingly, a hallucination is a percept not based on the material reality. On the other hand, as we have indicated, according to both scientific and anecdotal evidence, matter is far from being objective; in particular, it exists only when observed. Moreover, its observable existence and its properties are tentative and configurable by focused action of mind. Therefore, it fails as an objective reference point. No perception is really based on the material reality. It follows that everything that we perceive is a hallucination according to the original definition of hallucination.

The opposite of the claim that genuine perception is based on matter is true — the material world is based on perception (that includes imagination). As per String Theory, it appears that matter and its properties (including gravitational pull) are defined by the frequency of the vibration of the strings that constitute the units of matter. These can be influenced by action of consciousnesses. For example, true travel from "point" X to "point" Y may be no more than a change of the frequency associated to point X to that of point Y. Visualizing this in 3-dimensional space may be off the mark; everything happens within the inner domain of the undefined Supreme Consciousness.

Chapter 9

My Hallucinations

Deities have appeared in forms so visible that they have compelled every one who is not sense-less, or hardened in impiety, to confess the presence of the Gods.

Cicero, *The Nature of the Gods*

9.1 How It Happened

As we have noted in the preceding section, there existed initiates of high order who possessed the secrets of superimposing hallucinations on top of *normal* reality. Going in the opposite direction, there were prolific hallucinators who perceived *many hallucinated* scenes, possibly brought about by entities existing in *higher* planes. For example, according to the ancient Roman writings,[1] a certain Eucrates, who lived in the first century BC *has seen such spirits [apparitions] a thousand times, and from long habit, has lost all fear of them.* The Italian polymath, Girolamo Cardano (Jerome Cardan), 1501–1578, was another prolific hallucinator. In Section 2.4 we quoted him regarding his hallucinations.

I am far from being initiated into the secrets of hallucinations. However, I did my best to experience hallucinations, and at some point, I joined Eucrates and Cardano and became a prolific hallucinator. Ever since I *saw*, with open eyes and fully awake, the ghost of a woman in my bedroom in April 2007 (discussed in Section 2.4), I have tried to replicate the circumstances of that encounter and to get similar results. Eventually I managed to perceive, always with opened eyes, hundreds of apparitions, 573 of them deemed sufficiently interesting to be jotted down in my files. It should be patently evident by now that I vehemently disagree with the thesis that all of them are emanations from my brain. So, I considered them seriously and I tried to interpret them, especially those few that turned out to have veridical elements.

As far as I know, there exists no study regarding the frequency of prolific healthy hallucinators. My own small sample seems to indicate that such cases are not exceptional; out of no more than five people with whom I had discussed this topic, it turned out that one of them was a frequent hallucinator.

The working thesis during my trek into the world of apparitions was that the source of hallucinations was beyond the control of the bodily confined consciousnesses that we call humans. So, my attempts to induce hallucinations were combined with what can be described as my own efforts to reach out to the invisible powers and request assistance. My requests were sincere and made resolutely and with conviction. At times I meditated briefly before making such requests. However, there was no purposeful religious underpinning, and I did not pray.

I started by devising a simple procedure. As I mentioned earlier, when I saw the ghost by my bookshelves at the foot of my bed (Section 2.4), I was lying on my back with my head reclined on my pillow and in an almost vertical position. This is slightly uncomfortable, and one is bound to at least change the position after a while. I mimicked this position and learned how to open my eyes directly from sleeping. In most of my cases of hallucinations this was my modus operandi.

Eventually hallucinations started coming irrespective of how I lay on my bed. For example, in the second case (that happened about a month after the first one), I was lying on my side, and when I opened my eyes there was a smiling, bearded head of a man right in front of me. He looked to me like an ancient Sumerian or Assyrian warrior. The warrior never came back.

It is important that the perspective hallucinator condition himself/herself for the visions by learning to open the eyes directly from sleeping. Opening the eyes when awake seems to rarely lead to perceiving apparitions. I have one exception: once the apparition appeared **before** I fell asleep. In February 2014, I opened my eyes and I saw some dark spot in the direction of a window. I looked at it indifferently, expecting it to merge with a background. But it did not; instead, it started flapping what appeared to be wings, took the shape of a large black butterfly and flew into the wall, vanishing.

There were periods when the hallucinations happened whenever I requested them. Sometimes I voiced my requests audibly, but more often I did that silently, in my mind. After experiencing a hallucination, I usually expressed my thanks to the higher entities.

Most of the early apparitions were experienced as abstract (geometrical) objects, and some of them were fussy and unclear. There were also cases of extremely clear visions. Quite a few of them were apparitions of human forms, and I paid special attention to them; some such cases were clearly related to my current life. Most of my hallucinations were hypnagogic, experienced after the early stages of sleep; there were a few hypnopompic hallucinations, happening before one awakens.

Eventually I amassed several hundred cases of apparitions, and, as I mentioned earlier, I took notes on almost 600 of them. In most of the cases the apparitions dissipated as soon as I saw them. On a few occasions they lasted more than five seconds. At times the apparitions were very clear and so *real* that it seemed a gross misnomer to even call them "hallucinations."

Most of my early hallucinations were bizarre. For example, the fourth hallucination I saw, always with open eyes and fully awake, was a miniature human, a fairy of a sort, with red-painted face and normal proportions. It immediately glided upward and dissolved (May 2007).

For the fifth apparition (June 2007) I prepared myself by briefly meditating and asked verbally for a vision to come that night. About 45 minutes later (I checked the time) it happened! I was awoken by a humming sound, as if many uniformly rotating rods or whips were fast slicing the air. When I opened my eyes there was *something* very close to my head. It moved away and faded in the air. The something was opaque, round-ish and not more than one meter high. I have no doubt whatsoever that I had perceived it, that it was in a way an objective phenomenon. I tried to stay calm, but I could not control my excitement. I noted that I needed to control my excitement, which I eventually managed to accomplish.

In what follows in this chapter I will describe around 180 scenes from the fantastic world of hallucinations. It is supposed to be a casual reading. However, it is not fiction — my experience was as real to me as this computer in front of me. As mentioned earlier, I do not buy at all the thesis that, in general, hallucinations are exteriorizations of brain–produced thoughts. I strongly believe that in most of the cases hallucinations are brought about by sentient beings from higher planes of existence, or from higher frequencies. We have already established that it is at least a reasonable proposition to claim that such beings exist.

Here we go.

9.2 A Gallery of Hallucinations: Abstract and Ordinary Objects

I will continue to indicate the approximate time of each case so that we can position the events in my line of time, and so that we delineate more clearly the cases in my narrative.

During the early stages of my adventure there happened a long sequence of various hallucinations of objects or geometrical forms. For example, once (February 2008) I saw a spherical object with some undulations and protrusions. It did not dissipate, as in most of the cases, but rather went through the wall and out of my sight. There were also hallucinations of various bizarre objects, that opened up the unsolvable enigma of reasons why they have been shown to me in the first place. Here is a list of some of the hallucinated objects:

- A vine with grapes (April 2008).
- Chandeliers (many times).
- Flowers, single or in bouquets (many times).
- Other plants (many times).
- Crystals (many times).
- Christmas trees (several times).
- Balloons (several times).
- A pentaflake (pentagonal fractal; January 2009).
- A three-dimensional fractal (October 2008).
- A golden mask (October 2009).
- A pendulum (August 2010).
- A boomerang (December 2010).
- A whirling cloud of purple points (March 2011).
- Clusters of colored spheres arranged in various geometrical shapes (many times).
- A golden plate engraved with a picture of a man in a space suit (September 2012).
- A very bright object, almost emanating light, with concave hyperbolic walls and edges in a reddish-golden color, and with beaded corners (September 2012).
- A cluster of colored crystals (November 2012).
- A bike (October 2013).
- A large wire-meshed helicopter model; its blades rotated and the helicopter floated in the air above my bed (October 2013).
- A cross made of two bright-red painted knives (November 2014).

- A very clear view of a two-dimensional pattern that looked like a butterfly wing with Voronoy diagrams. It moved to my left and came very close to me; I extended my left arm to touch it (!), and it vanished (June 2016).
- I saw a crystal cup, perfectly cylindrical except for a small handle. It was posted on top of a crystal platform, and the whole contraption was slowly rotating (March 2018).
- Another time I saw, with my eyes widely opened, a huge crystal, positioned upright and around one meter tall. It was glowing yellowish light. It lasted the usual interval of time, 3 seconds or so (February 2014).
- Once there appeared an irregularly shaped rock that flew over my head; one side of it was fiery red, as if burning from inside (February 2014).
- A marble door handle, textured with a jaguar-skin pattern (June 2014).
- A small sailboat, seemingly made of straw (August 2014).
- A cylindrical object that tapered off at the top side into a ridge or an edge; the texture was polka-dot, with purple dots. It moved in the direction of the ceiling and sank into it (November 2015).
- A basket, made of woven twigs, and with a very long handle. Three nights later I saw the *same* basket, but that time I saw what it contained: flowers (April 2016).
- A small wooden elephant, decorated with precious stones, spiraling down in my direction. It hovered for a moment, then moved straight up, vanishing in the process (August 2016).
- A white vase with white blooming flowers. The vase was connected with wires to a hook above it; the bottom of the vase contained short semi-spherical legs. The vase went

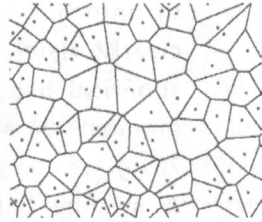

up into the ceiling and slowly disappeared, remaining clearly visible all the time; so much so that the semi-spherical legs, the last to merge with the ceiling, one moment showed as bumps on the ceiling (December 2016).

- A strange hallucination tonight: there was an object straight in front of me, but I was distracted by another thing that inserted itself between me and the object. So, I paid attention to the "intruder" and lost sight of the first object. The second object was a vine, or a stick covered with buds, extending from the bookshelf by my bed. I saw it clearly, since it lasted a good 3 seconds or so (January 2017).
- A shamanic medicine wheel consisting of a circular frame with eagle feathers all around it (November 2014).
- A thick silvery fork that flew close by the ceiling (December 2014).
- A vase, with a leaf-less plant in it, floating in the air right above me (January 2015).
- An ordinary snowman, complete with a carrot nose and coal eyes. It levitated in the air above me (March 2015).
- A small, rather rectangular, racing car, of an old brand; it was fixed in the space above me, but it shook as if it was moving along an uneven potholed road (March 2015).
- A full-sized motorbike, floating over my bed, to the right of me (August 2015).
- A replica of the tower of Babel! It was a clear sight and it lasted long enough for me to know what was going on, and to scrutinize it for a moment. I have checked images of the tower on the web, and none correspond to my

vision; "my tower" was more cylindrical compared to the artwork I have seen, where the tower is usually depicted as truncated cones (October 2015).

- Rings positioned along a fixed sphere, with threads of thin wires emanating from them and going in the direction of the center of the sphere. The central position was occupied by a small model of an exquisite sailing ship, and the wires seem to have supported it at its central position (January 2016).

- A very clear sight of a model of a pitch-black car from the early twentieth century, wrapped in transparent plastic (June 2019).

- I took a good look at the apparition tonight; it was moving obliquely from above the middle of my bed in the direction of my head, so the view was from many angles. It was a bulky old-fashioned camera. When it stopped in front of my eyes it was pointing the lenses in my direction. So bizarre! (April 2020).

- Two pages with handwritten text, suspended vertically, and pierced with two arrows, as if nailed on a wall. I couldn't read the text, even though the vision was very clear (October 2020).

- When I opened my eyes, there was a letter posted on the wall. Again I couldn't read the writing. It vanished promptly (October 2021).

- One night, instead of hallucinated objects entering "my world," I entered another world! When I opened my eyes I saw, very clearly, a structure from inside. The ceiling was replaced with a pyramidal roof, and the walls and the roof contained outlines of crisscrossing wooden beams. The vision was perfectly clear and lasted long enough — perhaps four seconds — to eliminate any doubts that I had perceived it. What was it? (May 2016).

- A weak apparition in the form of a rectangular object, close to the bookshelves on the opposite side of the room, that lasted not more than 2 seconds, and promptly dissolved into thin air. What was slightly unusual in this case is that I got a distinct impression that the objected emanated from my mind — that it was not created by consciousnesses external to mine (January 2017).
- Once I saw an open gate, exquisitely carved: a portal of a sort. It was suspended over my bed, and it looked like a small model, perhaps half-a-meter tall. I reached with my right hand to get hold of it — a non-spiritual reaction for sure — and my hand went through it with no resistance whatsoever. At the same moment the apparition disappeared (May 2009).

9.3 A Gallery of Hallucinations: Flying Saucers

Quite a few times I perceived miniature flying saucers or spaceships.

- Once there appeared such a miniature "spaceship." It came very close to my face, almost touching me. One moment it seemed as if it was going to enter me through my head. That startled me a bit (January 2014).
- Perhaps the same mini flying saucer, moving fast (November 2014).
- Another time there were three apparitions during the same night, all in the form of small flying saucers, with tiny yellow-lighted windows. The last of the three was the most substantial; it hovered above my bed for a couple of seconds, then spiraled away and vanished (November 2015).
- A very clear apparition tonight! There was a flying object above the foot of my bed. It was in the shape of a semi-

ellipsoid, oblong at the top, flat at the bottom. It flew directly in my direction, then abruptly changed its course, and went vertically up and into the ceiling (October 2017).

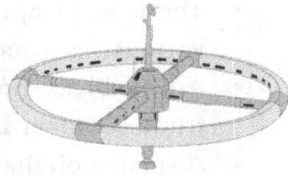

- A flying object in the shape of a ring connected with three symmetrically positioned spikes leading to the central sphere (September 2018).

9.4 A Gallery of Hallucinations: Animals

There were quite a few cases when the apparitions took animal forms — a whole menagerie of them.

- There was, for example, an apparition in the form of a silvery statuette of a horse, propped on its hind legs. It was amazingly clear, and that confused me for a moment, for I could not believe what I was seeing. Another feature was the way it disappeared; it did not fade away, but rather vanished in an instant (June 2012).
- Another time a rat dashed in my direction, then abruptly stopped, and vanished (June 2012).
- I once saw a white cat, also dashing towards me (July 2012).
- Then there was a fish with short spikes on its skin; it faced me directly and made moves as if swimming in the air in my direction (October 2012).
- A black fuzzy blob appeared at arm-length by the edge of the bed. I was calm; the object seemed to clarify somewhat in the form of an ape head with glowing yellow eyes. The ape had somewhat gentler features than a gorilla, with smaller cheeks and nose (November 2012).

- There were apparitions of a wiener dog, and cats of various sizes and colors (December 2012).
- A little black monkey climbing up and down an invisible trunk (October 2013).
- A small wolf that opened its mouth and showed two rows of canine teeth (October 2013).
- A reindeer head, equipped with large antlers (January 2014).
- A birdlike figure, with wings folded; it was rather a dull gray in color except for the beak that was perceived as red. It levitated above my head and vanished in degrees (May 2014).
- An orange seal (September 2014).
- Two different dinosaurs appeared during two consecutive nights. How much more bizarre this affair could be? The first dinosaur-like creature was perceived from its neck up. Its mouth was half open and there were teeth; not sharp but rather flat as in plant grazers. The second one was seen whole, and it looked like a two-legged bird-like dinosaur (pterodactyl?), except that his jaws were rounded and it had no beak (September 2014).
- A bizarre and short apparition, substantial enough to note it. I saw a cartoonish elephant, something like Dumbo, seated and facing me. It lasted two seconds or so, and it vanished fast (September 2016).
- Bizarre! There it was, facing me, a cat dressed like a musketeer, complete with boots, as in the picture book from my childhood (December 2017).
- Another bizarre cartoonish character in the shape of Bugs Bunny (September 2018).

- Yet another cartoonish character tonight: a small rhino, front legs resting on my bed, head leaning over them, winking at me with a happy face (July 2019).
- A huge snake, moving forward by lateral undulating moves, approaching me from the corner of the ceiling. It startled me for a moment, and it fast vanished (July 2017).
- I saw a lamp with a large snake coiled around it and facing me. I tried to touch it; it disappeared instantly (July 2018).
- Perfectly clear sight of a squirrel perched on its hind legs, hovering just above my head (June 2019).
- I saw the mouth of a sea creature: it was round with little polyps sticking out, their tops moving in circles. The unusual feature of the vision was that it was of very high resolution (May 2020).

Continuing the apparitions of animal forms:

- A large, white, fluffy dog, in sitting position, floating above the center of my bed. It did not move, and it did not vanish immediately. It stayed for a few long seconds, a sufficiently long time to fully perceive, then it started to shrink and disappeared in that manner (April 2015).
- A gigantic lizard, perhaps a crocodile, extending well above the ceiling. The ceiling did not exist in that scene, it was an open space above me (July 2015).
- Another time there appeared a smallish crocodile. It approached me rather fast, so there was a reason to be at least startled. However, I looked at it calmly — I was becoming a seasoned hallucinator. It came very close to my face and then it vanished (August 2015).

- A polar bear was the second apparition during the same night (August 2015).
- A kind of elliptically shaped animal, with a clear turtle-like head protruding at its front. It slid about a meter above the floor, at a steady pace and in a linear trajectory going just to the right of me (July 2015).
- A huge spider moving jerkily on the ceiling of my bedroom (August 2015).
- A large black spider appeared on the door of the closet. It started rolling in my direction and it dissipated as it came close to my head (February 2019).
- A full-sized bear head appeared not more than 50 cm in front of me. It was stationary and its eyes were fixed on me (July 2017).
- A cat! I was meditating, then dozed off. One moment I felt a cat jumping on my lap. I opened my eyes and saw ... a cat! It was standing on my lap. I did not know where I was; as I was trying to figure out my coordinates, the cat vanished! Strange case (August 2018).
- I saw, not more than 30 cm away from my head, a solidly golden (statue of a) wolf (or similar canine), motionless, head lowered as if to keep a low profile, gazing at distance as if observing a prey (January 2019).
- A funny looking orange snake with oversized head burst through a corner on the ceiling into my view. It made me smile (August 2019).
- In his book *The Sacred Promise*, Gary E. Schwartz recounted his "encounter" with his guardian angel.[2] Before sleeping he requested that his guardian angel manifest to him, and he got a hypnagogic hallucination, in the form of an 8-foot-tall blond, beautiful woman who emanated powerful light! So, I tried the same: going to

sleep I asked the divine entities who hear me to show me my own guardian angel, if I have one at all. I got it, but it was not a beautiful angelic woman. What I got was a monkey, peacefully sitting on the bed by my side. The entity was not illuminated at all, and it disappeared in the usual way. I am still smiling; instead of a beautiful blonde woman I got a monkey! (May 2022).

9.5 A Gallery of Hallucinations: Humans and Humanoids

At the beginning of my adventures into the world of hallucinations most of the perceived objects were inanimate or abstract. Eventually human and humanoid apparitions begin to dominate, so that there is an abundance of such cases.

- *Something* made me open my eyes and I had a very clear vision of a humanoid *entity*. It lasted about five seconds, long enough to get clear impressions. The entity was between my bed and the windows, apparently *standing* on the floor. It was human-looking but somewhat smaller. The impression was that it (he/she) was a *female* (if it makes any sense to use gender in these matters). The creature did not advance in my direction; it just stood still. I think it (she) had her hands stretched in my direction. My analytic mind took hold of me, and, *knowing* what the entity was, I wondered if it would then begin dissolving. The moment this came to my mind, the entity fast dissolved into thin air (April 2008).
- I saw a gallery of human-looking entities: ordinarily looking human forms, both male and female (many times).
- A turbaned, bearded man (April 2008).

- Many cases of apparitions of children, including a boy with curly blond hair (September 2014).
- Many cases of miniature human-like entities; a small golden statuette of a woman sitting on a stone throne (May 2012).
- A warrior with a mustache like an upside-down U, dressed in some oriental costume (May 2012).
- A hatless straw (wo)man (September 2012).
- I clearly saw the head of an Indigenous American, with hair parted in the middle, and a stern, cold face expression. It was in semi-profile, with prominent, somewhat hawkish nose and protruding cheekbones (June 2013).
- I also noted in my files the apparition in the form of a woman seated on a stool beside my bed. She wore long skirt, covering her legs all the way to the ground. There was nothing hazy or unclear about the woman; she appeared completely materialized with minute details. The only problem — and major problem, for that matter — is that the *vision* lasted but a single moment, as if there was not enough energy, or power within me, to sustain it for longer (October 2012).
- There followed the apparition of a little black man — black in the sense that everything on him, clothes, and skin, was tinted black. Perhaps less than a meter high, he came about as a small copy of Don Quixote, with medieval attire (September 2011).
- The bust of a man that looked to have been gypsum-made, positioned horizontally above and a bit to the right of me, and facing me (September 2013).
- During two consecutive nights there appeared very close to me a dwarfish little guy dressed in a red cape; he/she/it disappeared in the usual manner (May 2014).
- A few nights later I saw a man standing at the foot of the bed. The image lasted about two seconds and dissipated

rather fast; so, I could not see much. It appeared that he was wearing a soldier outfit, perhaps some khaki overalls (May 2014).

- There followed a doll-like large holographic image of a woman, illuminated with light and encased in something that looked like a wooden box. The face was expressionless, and there was no movement or grimaces of any sort; the whole scene shifted in the direction of the wall behind me and disappeared (September 2014).

- I clearly saw in profile a middle-aged man. He was dressed in white robes of some sort, perhaps a tunic. His nose was straight, his hair was carefully combed backwards, and it was completely gray. He gave an impression of a wise, stately person. The vision lasted a few good seconds and then it vanished (June 2016).

- A girl in a red conical skirt appeared very close to me; her head was the first to disappear, and then the whole trunk went out of perceptual existence as if being turned off (July 2016).

- I saw a bold head, or a skull, positioned on top of a rectangular body. It was looking away from me. It appeared to me rather robotic or metallic. It did not move. The head was the first to vanish, and then the "body" merged with the door at the background (July 2016).

- Two human apparitions. The first was a man in his thirties or so, with a disheveled countenance. He looked at me without moving. The second one was a woman in some robotic, angular attire. Her hand was stretched in my direction and her index finger was pointing at me. This vision lasted long enough for me to request that she talked to me; I am not sure if I actually said it, or I only made the request in my mind. The vision dissolved, with parts of it merging with the white bookshelves in the background (July 2016).

- When I opened my eyes, I saw a figure of a man in profile; he wore a hooded overall, like a monk. He was with his hands close to his face, appearing to put a mask of some sort on his face (April 2017).
- I saw (with open eyes, of course) a feminine apparition, about a foot long, leaning in my direction and not more than a meter away from my eyes. She wore translucent, burlap-like clothes of ancient times. As soon as I open my eyes, she (it?) pulled back and straightened up, then vanished (May 2017).
- A man in a seating posture, facing me. He was dressed in colorful clothes, almost clown-like. The apparition lasted some five seconds. I said "hi!" and the being vanished (July 2018).
- A boxer. Fully outfitted, and with red boxing gloves. The posture was classical: semi profile, hands bent in front of his body, gloves at chest level. He looked serious as he slid from the foot of my bed towards my face without changing its posture. Very clear and relatively long-lasting apparition (August 2018).
- A man in his thirties, who came on three different occasions. He was clean-shaven and good looking, with a straight nose (not a large nose). For some unclear reason he reminded me of myself (May 2020).
- The head of a woman was right in front my eyes. She had a three-crested hat of some sort, and a flat large nose to match the crests. The apparition was very clear but lasted not more than three seconds. As soon as it vanished, the

light that surrounded the head disappeared too and I was in dark again (August 2022).

- A middle-aged woman talking animatedly and somewhat angrily, pointing her finger towards her mouth. (What was she trying to tell me?) The apparition was clear and lasted some four seconds. Later in the night I saw a man with a fedora hat; he was also pointing at his mouth (August 2022).
- There was a man of about 30, standing by the foot of my bed. He was looking away from me and gradually turned his head in my direction. The moment our eyes made a contact he disappeared (August 2022).
- A woman was lying in the bed to the left of me; her arm was around my chest. She turned around toward the edge of the bed, and, as she was lying on her left side, she sank into the bed or over the edge (September 2022).
- A middle-aged woman who looked like a housewife of the 1960s, with a matching hairdo. Later that night I saw a handsome man in his thirties (September 2022).

More human or human-like apparitions:

- A human contour, made of a discontinuous mesh of wires or three-dimensional curves (May 2015).
- A woman, dressed in a white wedding gown, slightly inclined in my direction (so that she was perpendicular to my direct vision). I thought she was going to jump on me — which startled me. Her arms up to her elbows were stretched out, bended at a right angle at the elbows, pointing up. She moved in my direction as she waved happily with her hands (June 2015).

- A friendly-looking, smiling woman (August 2015).
- A girl dressed in a white uniform, smiling at me (July 2019).
- A little sad girl of about 5, carrying a jug. Her face was very clear; it reminded me of someone I had seen (April 2019).
- A woman appeared by the side of my bed. She was dressed in a simple, white neck-to-ankles wardrobe, similar to Slavic women's dresses of Russia or White Russia. Her arms were outstretched holding a thin, short object, like a feather. The moment I realized what I was seeing my heart started beating faster, and the vision immediately vanished (October 2018).
- A boy of about 2 years, standing at the foot of my bed. One of his hands was raised, and his head was leaning on the palm. He looked at me calmly and with a bit of curiosity (May 2019).
- A blond girl of about 6, holding a toy car and walking (May 2019).
- A woman standing with outstretched arms, holding something like a stuffed teddy bear. She approached me and raised her hands as if offering it to me as a gift. I extended my arms to take the gift, merging with the gift with no resistance; the vision vanished immediately. The same woman appeared the following night. She talked to me, but I couldn't hear anything. It appeared as if she was scolding me. I asked her who she was, and the vision disappeared (August 2019).
- A sad woman seen in profile, resignedly throwing a pill into her mouth (December 2019).
- A very clear view of a middle-aged man with a short, well-kept beard; it seemed as if he was waiting for me to see him. He looked at me sternly (March 2020).

- There were three separate apparitions. The first one was a view of two soldiers: one was with a De Gaul military cap, the other carried a gun over his shoulder. Then there was an ordinary looking man in his thirties. The climax came with the last apparition. There were two men again, one seemingly seated on the bottom of a narrow frame of a window, the other, the man from the previous apparition, facing him and apparently talking to him! There was no sound that I could hear. One moment the second man turned his head in my direction, and the vision disappeared (June 2020).
- The wall in front of my eyes was like a big movie screen. I saw an oriental guy, about 30 years old, with a motorcycle helmet, approaching a camcorder or similar posted on a tripod. He was about to look through it, when the whole scene vanished. How bizarre! (June 2020).
- An Asiatic woman bending over me, her upper body straight; when I opened my eyes she sprang into vertical position, keeping her spine straight all the time (June 2020).
- A middle-aged man, looking like a tramp, with an unkept beard going in all directions, grinning rascally at me, and launching forward as if attempting to frighten me (July 2020).
- A doll in a fluffy red skirt, standing upright and rotating or pirouetting (August 2020).
- A little girl of a year or so standing in the middle of my bed. She wore a wool-cap of the sort. She stood motionless for about five seconds (August 2020).

- A blond doll-like girl floated above me; it was a pretty clear sight (September 2020).
- There was nothing fuzzy about tonight's apparition; it was the same as seeing under waking consciousness. A blondish man, with a headband, and T–shirt revealing his bodybuilder muscles, stood beside my bed, partially turned away from me. I was bewildered; I couldn't make sense out of it. It lasted about five seconds (September 2020).
- First there was a middle-aged woman standing by the foot of my bed. She was pointing (something; a gun?) at me, smiling pleasantly — but perhaps with an "I got you" smile — and looking directly at my eyes. She was talking nonstop, but I couldn't hear a sound. I smiled back and she disappeared. Soon after I saw the head of a bearded man; it was motionless and looked like a sculpture. The face reminded me of myself, but it was more angular. The beard was undulating, but not very long, more Greek than Assyrian (November 2020).
- I saw (with open eyes, of course) a very pretty young woman, standing by the foot of my bed. She wore blue jeans and had short brown hair. She was looking at me sideways, smiling sweetly. I hope she returns [she never did] (December 2020).
- I saw a woman; she was holding something in her half-outstretched hand, and she was looking at me with a conspiratorial smile, as if teasing me (February 2021).
- A boy of about 10, facing me. A few minutes later the same boy appeared; he was looking at me. In both cases the vision dissipated slowly (March 2021).
- So bizarre and so amazing! A feminine being that looked like a scarecrow — extremely clear and very real. Her head was made of burlap, with cartoonish eyes, and a mouth made of buttons or similar; smiling at me and

leaning forward as if checking
if my eyes are open. When she
saw that my eyes were open she
looked satisfied, leaned back,
stood up and disappeared
promptly (May 2021).

- A woman sitting by my side.
  She wore ancient mid-eastern
  clothes, and she had a small
  baby on her lap. She was looking at the distance, without
  paying attention to me. The scene reminded me of the
  classical depiction of Mary, the mother of Christ. Very
  clear vision but lasted not more than 3 seconds (June
  2021).

- Tonight's apparition lasted around 10 seconds. I saw a
  stern-looking black man who reminded me of the actor
  Morgan Freeman. There was a hint of a smile on his
  face; something like a "what-now-smile." I smiled back
  hesitantly. I felt uncomfortable; it seemed to me he was
  silently criticizing me. I asked him, "who are you?" He
  didn't move; then he vanished (August 2021).

- A man in profile, nodding his head up-down, as if calling
  me to follow him. I said OK and he instantly vanished.
  Later the same character appeared once again (September
  2021).

- A young Asian man with short beard and black-framed
  glasses, standing in profile. A rather unusual feature was
  that his face was illuminated (October 2021).

- A red–haired, red–bearded angry young man, talking
  fast and waving his hands. He was completely inaudible
  to me (May 2022).

- Gentle feminine hands close to me. The vision was very
  clear, so the first moment I was not sure if what I was
  seeing was not a "real" thing. When I realized it was a

"hallucination," I looked at it a few more moments, then extended my hands in its direction, as if trying to hold hands with it. It vanished instantly, and my hand went through the "empty space" (July 2022).

- A little girl of about 4 appeared at the foot of my bed. Her hair was straight and relatively short. She had her back towards me. It was not a clear apparition, and it lasted two seconds or so. The same little girl showed up again the next time I opened my eyes; she was climbing onto my bed (May 2023).

- I saw a wooden gnome or elf, about 30 cm tall, floating in the air in front of me and about 50 cm above me. It was neck-less and as if made of a single log. The creature was not animated; it looked like a rough wooden doll. It moved along a concave up parabola, and as it started going up, it vanished. It was very clear at one point, and it lasted long enough for me to realize what was going on while it was still visible (March 2016).

- A long-lasting apparition tonight — more than 5 seconds! It was a round, doll-like creature, that stared at me from above the middle of my bed. It was yellowish, with rounded contour, like an owl, but with a humanoid face. We stared at each other motionless for a while, then it cocked its head on one side, and melted away. The length of the vision may have been the result of my calmness (June 2017).

- I asked for a human apparition tonight, and I got one: Woody! How bizarre! A doll-like creature, with a cowboy hat, a checkered shirt and a vest; it was about a meter tall, as it stood right by my bed (July 2017).

- A dwarfish human figure popped up in front of my face. I only noticed its head; it had waxed texture, and it looked like a version of Popeye! Its eyes were moving; after a few seconds it slid sideways and it vanished (August 2017).
- I clearly saw a woman at the foot of my bed. She was standing and seemingly pointing at something. The woman wore jeans and was relatively young (30s?). This would not have been anything extraordinary except for the fact that the same woman appeared two more times that early night. I don't have visual recollection of her second "coming"; the third time she was by my bed, she was again trying to show me something (October 2017).
- A hip of white–painted human bones, with the skull dominating the view floated above my bed. The skull fell down and the scene vanished (April 2018).
- A round-faced plump man in monk's attire, walking in the direction of the wall to the right of me, seemingly in deep thought and not paying attention to me. During the same night I also saw a girl with her face covered by a cat-mask; her head was cocked sideways and looking at me, as if playing peek-a-boo (August 2019).
- There was a small girl holding a doll and standing by the foot of my bed. She was rather clear but disappeared promptly (October 2021).
- The vision of an unknown middle-aged man; he was very close to me, as if sitting on my belly. The apparition ended with the guy smiling at me for one single moment. Soon after the same guy appeared in profile further away from me (November 2021).

- I saw a statue of a headless woman wrapped in an ancient flowing garment. The apparition looked like a perfect sculpture made of white marble; it lasted about three seconds (December 2021).
- A black man was standing by the foot of my bed, sideways with respect to my position. His head was facing in my direction and he was pointing at me (January 2022).
- A boy of about 6 appeared twice: first at the foot of my bed and then above the center. He was dressed in white, loose, silky robes (January 2022).
- I saw a hand, with fingers stretched upward. It was a white, cloudy texture, a proverbial ghost, and it was much smaller than an adult human hand. Actually, it looked like a hand of a human embryo, made somewhat translucent by a kind of light in the background. The vision lasted a second or so, long enough for me to instinctively trying to touch the hand (June 2016).
- A strange case tonight: I dreamed of my mother; within the dream I realized I could see her if I open my eyes. So I did. And I saw her momentarily — she was seated on the chair by the foot of my bed. She was looking down and to the side. The vision lasted one moment so I cannot be sure. But it was not a dream (April 2020).

9.6 A Gallery of Hallucinations: Noticing Me!

In a few of the previous cases the apparitions reacted to my presence. In cases of human apparitions, this mostly consisted of looking at me, pointing at me, approaching me, talking to me, trying to show me something, and sometimes turning the head so that it faced me. More such cases follow.

- Once a little, dark, doll-like creature approached me, which startled me, which in turn caused the little man to disappear (June 2017).
- Another time, when I opened my eyes in the hypnagogic state I saw a man; he was by the bookshelf by the closet appearing to be seated and facing the wall behind me, so that I initially saw him in semi-profile. He had a medium-short beard and a mustache and gave an impression of a monk of some sort. He turned towards me, and pointed a finger in my direction, looking at me sternly and seriously. The scene lasted about five seconds — a long time (September 2013).
- In a similar case, after I asked eagerly for a human apparition and the Great Benevolence kindly obliged, I saw a woman in all white, lying on the floor a meter away from my bed, parallel to me. She turned to face me one moment (October 2013).
- An unusual case of double-apparition happened during another night. It started with a view over a black-and-white small hospital monitor, and then the scene abruptly changed into a drab human-like entity that stretched its hands and lunged upon me. That startled me (October 2013).
- A clear, long-lasting, interactive apparition! There was what appeared as a holographic projection of a woman standing above the middle of my bed. The image was about 30 cm (or a foot) in height. She wore a monk wardrobe, a hooded coverall with wide sleeves. The hood was not obscuring her face — even a strand of hair was visible. We looked at each other for a few moments, and then she actually walked — or moved her legs as if walking — in my direction, stopping at arm's length from my head. Her right hand was extended, holding something I did not see, the other was bent at the elbow. She started talking — but I heard no voice. Her demeanor

was serious, even somber — and she came about as if scolding me. She seemed to be saying, "what are you waiting for?" Then she stared at me without moving. We looked at each other for a few seconds. I extended my hand in her direction, and she vanished (February 2018).

- A middle-aged woman right beside me; she was talking to me in an agitated manner. No sound could be heard (which in my case does not mean much). Lasted about four seconds (July 2022).

- A woman strolled from my right to my left above the center of the bed, as I was lying in the usual position conductive for inducing or perceiving apparitions. She walked casually, while at the same time talking to a child (which I have not noticed), or to someone much shorter than her. I smiled as I was watching the scene. Unexpectedly, she turned toward me, raised her head, and smiled back (April 2020).

9.7 A Gallery of Hallucinations: Auditory

A couple of times the apparitions were preceded by auditory hallucinations.

- Once I heard in my mind an order, or a request to wake up. The apparition that followed was a non-descript dark form hovering over my bed (May 2009).

- I got a funny auditory hallucination while meditating: a jovial yoooo-hoooo, loud and clear, straight into my head (August 2019).

9.8 A Gallery of Hallucinations: Scary

A warning to anyone trying hallucinations: some of them will undoubtedly be unpleasant, even scary. In such a case you may have the opportunity to abandon the experiment. However, there is a chance that you may end up with something you cannot control. It is the risk taken by any prospective hallucinator.

I experienced a few supposedly *scary* hallucinations; sometimes they made me reflect about the people I have hurt in my life, but it did not deter me from continuing with the experiment.

- Once I saw a zombie-like woman with a rotting face who stared at me fixedly. I simply stared back. I was not afraid, and my heart did not start beating fast. Since I was calm, the vision lasted somewhat longer: perhaps four seconds or so. Then it vanished (March 2012).
- Another time, a little dragon that approached me head on. I was startled one short moment, then immediately decided to be calm and observed it; this seemed to cause a change of the course of the apparition — it immediately withdrew, then disappeared (June 2012).
- And there was an apparition in the form of a huge black spider-like entity, suspended right in front of me. I said "stay!" and even tried to get hold of it. It pulled up — with a rather spider-like move, as if hanging on a thread — and it started vanishing at the same time (October 2012).
- The gallery of supposedly scary apparitions included the skeleton of an arm, extended in my direction so that the bones of its fingers were close to my face. At that time I was a seasoned hallucinator and this apparition did not phase me off (June 2013).
- When I opened my eyes there was a middle-aged woman standing on top of my bed and staring angrily at me. She was holding a stick, ending with three sharp prongs, like a trident. She raised the stick and thrust it straight into my eyes. I didn't even blink; I concentrated so that the

apparition lasted longer. I don't know what this episode means. Am I being tested, or does someone simply hate me (August 2019)?

- The head of an old (or dead) woman, not moving, lying directly on my stomach; it looked as if it had been severed from the rest of the body, and that startled me (April 2020).

- A "witch"! An unnaturally round, ugly, pink-skinned woman, with a face of a witch. She was positioned as if strolling by the foot of my bed, but her face was turned in my direction, looking at me with a sinister grin. Lasted a couple of seconds. I requested that she re-appeared, but she didn't return (November 2020).

Here is a note straight from my apparition cases: "I earnestly asked for an apparition tonight, and the higher entities obliged. I consider this affair to be a small window to alternative percepts, and I will do my best to (whine in order to) keep it open. However, I am not able to discern any deeper meaning from this whole affair other than the naked fact that the apparitions manifest. Hence, I take the phenomena casually — perhaps too casually — since I don't really try hard to memorize the details of the forms of the apparitions. This happened tonight. I saw an illuminated object in the corner of the ceiling by the door. It looked like a lantern. It moved slightly in my direction, and vanished. I murmured my thanks to the higher entities and went on with the business of sleeping." (March 2016.)

9.9 A Gallery of Hallucinations: Requested Objects or Humans

There were quite a few apparitions of people whom I have known. The triple encounter with what seems to have been the ghost of Timothy Wyllie was described in Section 3.3, and the three apparitions that were the most important to me are included in the Epilogue that follows.

A few times I perceived hallucinations related to the requests that I had made before falling asleep. For example, I once saw, very clearly, a red five-pointed star. It was suspended above the foot of my bed in a vertical position, then moved upward in a semi-circular trajectory reaching the top in horizontal position where it vanished. The context of this apparition seems significant: before relaxing to sleep I invoked Jesus Christ and requested that He manifests to me in some way. The five-pointed star, also known as the Star of Bethlehem, is a symbol in Christianity representing Jesus' birth (December 2018).

- Someone tapped my foot while I was sleeping. I opened my eyes, and I clearly saw a cute little brown-haired boy, looking directly in my eyes and smiling sweetly. This apparition happened while I was in a rented house in Calgary (September 2019).
- Two years later I stayed in the same rented house, and I requested that the boy appeared again! He obliged. His head was protruding from the wall by the foot of my bed. I decided to follow the cliché and wished him that he goes into the light. He disappeared. I wondered if the young boy had lived and died in the house (September 2021).
- A very clear snapshot of a man. He was bearded and half-naked, with wide rectangular shoulders, and with long, misty dark entangled hair. This scene was not animated. What is interesting about all of this, is that I requested before sleep to be shown an incarnation of "myself" far removed from the present. Was my request granted? (November 2020).
- Before I went to sleep, I asked the powers that be to let me see the previous incarnation of a girl dear to me. I got a

very clear apparition of a young, blondish boy reminding me of her. He moved slowly in my direction and promptly disappeared (October 2021).

9.10 A Gallery of Hallucinations: Vibrations

In a few rare cases the hallucinations were tactile, causing waves of vibrations. They carried a measure of significance, and perhaps influenced me physically in a tangible way. We provide a few examples:

- Once while I was asleep something, or someone, touched the arcs of my feet. It was a simultaneous and very strong touch, almost a push. It produced a surge of energy moving up along my legs into my torso (September 2014).
- Another time, as I lied on my back, I felt a distinctive "touch" on my legs, as if something put pressure on them. Then there was a huge wave of vibrations overwhelming my arms, and especially my legs. I was literally oscillating at a very high rate. I did not resist the wave, but I did feel agitated. This lasted perhaps six seconds (July 2016).
- Tonight's vision was exceptional! In itself it was nothing spectacular: an angular crocodile head, bounded by a very few rectangular polygons, robotic looking and moving smoothly along a curved trajectory. I looked at it calmly and with curiosity. Then something happened. As the contraption was moving above me, my legs started vibrating. At first it was gentle, like goosebumps; then it gradually increased and spread throughout my lower half of the body. At the end of the experience the amplitudes of the vibrations were so large that I was literally shaking. Amazing! What to make out of this? (May 2020).
- I saw a very tall human or humanoid with a relatively small head, standing by the foot of my bed and looking

at me. A few moments later the vibration experience repeated: it started with my legs and moved up, increasing in frequency and amplitude, while at the same time I felt increased warmth. The same repeated a few minutes later, but without the apparition (May 2020).

- A very interesting triple vision! An entity that looked like an oversized wooden doll appeared at the foot of my bed. It stared at me without moving, then vanished. At that moment I felt very strong vibrations over my lower body. I was literally shaking. The frequency of the vibrations was not very high, but they were of longish amplitudes. Sometime later, the "wooden doll" appeared again. After it vanished, I tried to simulate the vibrations in my mind. Amazingly, when I opened my eyes, there it was again, standing beside me, then taking one step in my direction, and disappearing (November 2020).

Vibrations of the body have at times been reported during the onset of near-death and other out of body experiences. Was I being guided in the direction of an OBE during the vibrations episodes? Perhaps I was too much attached to the *normal* plane of existence and was thereby not responsive enough.

9.11 Concluding Remarks

My adventures in the world of hallucinations ended with the following episode:

- I saw a child of about 3 years old being held sideways by a person, possibly another bigger child. The apparition was very brief, and I only had a glimpse of the smaller child. He didn't look like any of my kids or grandkids; in fact, the impression I got was that he was me. This was my last apparition. It happened on July 31, 2023, my birthday. Coincidence? I doubt.

What to make out of these cases? I feel I have failed to progress to a level from where I could understand them better. For example, I could not hear the apparitions, despite their repeated attempts to talk to me. In a few of the above cases my inability to hear them seems to have made them frustrated or even angry. Have I failed?

The culmination of my adventures in hallucinating happened in the two cases described in the next chapter.

Chapter 10

Epilogue

When the invisible becomes visible, you can never fall back into ignorance.
Anke Evertz's guide during her NDE.[1]

The following three "hallucinations" are the most important to me.

10.1 My Father and My Son

The apparition on 19 June 2023 was unique in many ways: I saw — with open eyes — my father holding my son in his arms. My father — deceased for seventeen years — appeared much younger than when he died. He looked healthy, and he was smiling and beaming happiness. My son was shown as a toddler of about 2 or 3, still with blond hair. My father's arms were slightly stretched in my direction, as if delivering my son to me. He was smiling radiantly; my son was observing me with curiosity. They were both facing me, and both were looking at me. The apparition was brief, but very clear!

I assumed the apparition was not a random act, but I could not discern a clear meaning. Then, as it seems, things clarified. A few days later I learned that about a week before I saw the apparition my son had had a serious incident: his heart almost stopped beating for a couple of minutes or so. He pulled out of it without any apparent consequence. Was he helped by the higher entities who control our lives? Was this the meaning of my apparition — was the delivery of my son by my father in the scene that I saw a metaphor for the delivery of my son's life during his heart episode by the invisible controllers of our lives? Did they steer my son into the calm waters of healthy life? It seems to me that the affirmative answer is a reasonable proposition.

I came to believe that this was the meaning of the apparition. I will be for ever grateful to the higher entities for their assumed intervention, and for actualizing a timeline away from what would have been a tragedy for our family.

There is a unique aspect of the apparition that needs to be noted: it concerns its source. To the extent the individual consciousnesses are separated, the apparent holographic image of my son as a toddler within the hallucinated scene was not a creation of my adult son from the perspective of our tentatively called present. The consequence of this statement and of my strong conviction that hallucinated scenes are, in general, not brain emanations, is that the hallucination of my father and my son was created or assisted by some entity from the higher planes of existence. This may always be the case with hallucinations; the apparition of my father with my son strongly supports this claim.

10.2 Tanya

All of the flowers blooming together, Tanya.

The above is an insert from a longer verse that I remembered from a dream that happened during the night of October 5–6, 2020; this came as a song, but I lost the melody.

The apparition on January 26, 2019, seemed like one of many that I have experienced. When I opened my eyes, I saw a pretty woman standing by my bedside and gazing in the direction of the center of the bedroom. She was smiling! The vision lasted a couple of seconds or so. I had no time to react.

What is significant to me about this simple vision is that, before falling to sleep that night, I silently requested that Tanya appeared to me in what would have been her body at the peak of her life. (I didn't expect anything, but I still made the request.) Just because of this *coincidence*, I knew immediately that there was a chance that the vision may not have been a random hallucination, that what I saw was indeed a representation of Tanya as a young woman.

I also noticed a certain peculiar feature regarding her face that I was not aware of. This was confirmed the next morning. To me this was a sufficient proof that it was a 3-dimensional image of Tanya that I saw. So, I remember well, and will always remember, the smiling face of the pretty woman.

The apparition of July 23, 2019, touched the bottom of my soul. At first the vision was blurry, but it immediately cleared into the features of a pretty woman. This time I recognized her instantly: Tanya! She was closer to me than arm's length. She stood, looking in my direction, and smiling happily. I said "I love you," she gave me a distant kiss, and vanished. The vision lasted around five seconds.

Smiles and tears.

Endnotes

Introduction
[1] [Ros]*, p.250–258.
[2] [Sco], p.42.

Preludes
[1] Name changed.
[2] Postscript: It is April 13, 2022, today. Yesterday I picked up a random book from the shelves of the university library. It is *The Quest for the Fourth Monkey* by Sylvia Fraser. I found the following on page 48: "Eventually I go to my room where I fall into heavy doze. 15 minutes later I am jolted to attention by an unearthly and unidentifiable shriek. I look at my watch: exactly 3 o'clock." It transpired that her father died at the same time, some 4000 kilometers (2700 miles) away. So, as I suspected, there exists a precedent to my experience. Very likely there are many such precedents.
[3] [Stra], p.262–263.
[4] This is now called a shared death experience (SDE), a phrase popularized by Dr Raymond Moody in his book *Glimpses of Eternity* ([Moo]). It is defined "as occurring when a person dies and a loved one, family member, friend, caregiver or bystander reports they have shared in the transition from life to death or have experienced the initial stages of entering an afterlife with the dying." See [Pete], p.5.

Chapter 1
[1] [Sac], p.xi.
[2] [Blo], p.14.
[3] [Grof2], p.216.

4 It is an interesting tidbit that the terminology "hypnopompic hallucination" was coined in 1901 by the physicist and psychic researcher, Frederic Myers, whom we will encounter several times in our journey.

5 [Ree], p.37.

6 [Das], pp.307–309. The original source is Ludwig Mayer's *Die Technik der Hypnose*.

7 "Ghost" is such a worn-out notion that we will avoid it.

8 Graham Reed: *The Psychology of Anomalous Experience*, p.55.

9 [Rin], p.165.

10 [Bud], p.105.

11 and future too!

12 [Eva], p.94.

13 Ibid., p.123.

14 [Jun], p.248.

15 [More], p.170.

16 By the way, More had his own classification of "ghost of dead men": he divided them into those appearing in dreams, and those perceived by "open vision" (during waking states of consciousness).

17 [More], p.173.

18 [Sac], p.234.

19 [McC], pp.288–289.

20 More than a million according to some estimates.

21 [McC], p.71.

22 Ibid., p.81.

23 [Bai], p.81.

24 [Ree] p.37.

25 [Blac], p.83.

26 Ibid., p.85; a table of surveys of the OBEs is given on page 88 of this book.

27 [McC], p.288–289.

28 [Doo], p.89, article by Rogo.

29 [Gug], p.12.
30 [Rad], p.89–90.
31 [Pop], pp.127–129.
32 [Osi], p.29.

Chapter 2

1 [Val], p.51.
2 [Rei], p.290.
3 [Bar], p.11.
4 [Mor], p.58.
5 The mechanism of tinnitus production is also controversial (American Scientific; https://www.scientificamerican. com/article/what-causes-ringing-in-th/).
6 See for example Grant Cameron's *Tune in: the Paranormal World of Music.*
7 [She], p.xx.
8 This is said without a slight intention of denigrating such composers; indeed, our main (ancient) thesis is that everything we perceive though our physical senses, including the perceiver, is a kind of hallucination.
9 [McC], p.338.
10 Jérôme Lalande, *Voyage d'un François en Italie.*
11 [Sac], p.108.
12 It seems that William Blake was the one who introduced that phrase: "If the doors of perception were cleansed, everything would appear to men as it is, infinite."
13 Note: since I do not subscribe at all to the linear, one-way oriented model of time, this sentence is tentative. Chapter 6 will be devoted to time.
14 [Mae], p.152.
15 [Lun], p.210.
16 [Mor2], p.126.
17 [Car].
18 [Mye].

¹⁹ http://members.aol.com/timeslip8888/.

²⁰ [Eva], p.57.

²¹ Not his real name.

²² http://www.users.csbsju.edu/~eknuth/pascal.html.

²³ [Stre], p.5.

²⁴ Ibid., p.6.

²⁵ [Smit], p.66.

²⁶ [Yag], p.28.

²⁷ [Carr], pp.205–221.

²⁸ Here is my own exception to that rule, as recorded in my files on June 8, 2021: "I *saw* a woman sitting on the bed by my side. She wore ancient mid-eastern clothes, and she had a small baby on her lap. She was looking at the distance, without paying attention to me. I recognized (?) her as being Mary, the mother of Christ. A very clear vision; however, it lasted not more than two seconds."

²⁹ [Chr], p.29.

³⁰ [Rad2], p.50.

³¹ [Bru], pp.68–72.

³² Grof's article in Gary Doore's (Editor) *What Survives* ([Doo]).

³³ Horatio Donkin will be introduced in the next section and in the endnote³⁵, this chapter.

³⁴ [Ose], p.116.

³⁵ Horatio Donkin was a physician and a part-time crusader against anything that contradicted the tenets of the orthodox science of his time. He once sued the powerful medium, Henry Slade, for cheating ("taking money under false pretenses"). The judge's ruling is instructive: he invoked "the known course of nature" and convicted Slade to three months of hard labor. So, Slade's challenge to "the known course of nature" through production of some amazing phenomena (see, for example, Zölner's book *Transcendental Physics*), was denied by invoking "the known course of nature"!

[36] https://www.abo.ru (in Russian) and https://www.abo.ru/english.html, in English, abridged version.

Chapter 3

[1] From Blake's poem *Auguries of Innocence*.

[2] [Fod], p.23.

[3] [Hil], pp.82–85.

[4] [Fod], p.82. The original source is *Journal of the Society of Psychical Research*, 1908.

[5] [Gra], p.141.

[6] [Oat].

[7] [Cro], p.25.

[8] [Ste], pp.147–148.

[9] [Sla]*.

[10] [Rich], p.49. The original source is *Histoires Prodigieuses* by François de Belleforest, Commingeois (published 1578).

[11] [Gug], p.12.

[12] [Fla], p.202.

[13] [Fla2], p.233.

[14] [Bai], pp.83–85.

[15] [Fla3], p.52.

[16] [Fla3], p.72.

[17] [Mar], p.10.

[18] [Smith], p.20.

[19] [Stea], pp.124-126.

[20] Also in [Kni], pp.137–139.

[21] [Har], p.28.

[22] [Cal], p.321.

[23] [Fla2], pp.141–143.

[24] [Jaf], p.95.

[25] [Wils], p.105.

[26] [Bar], p.15.

[27] [Kes], p.7.

²⁸ [Moo], p.16.

²⁹ [Fla2], p.313.

³⁰ Quotes around "survive" are in order to distance myself from the implied linearity of time. I will more often use italics when I want to distance myself from the prevailing meaning of the word, as, for example, *ghost*.

³¹ In fact, in its native state consciousness has nothing to do with (what we call) matter.

³² [Tro], p.163.

³³ [Wan], p.151.

Chapter 4

¹ "A cross-cultural study of beliefs in out-of-the-body experiences, waking and sleeping," *Journal of the Society for Psychical Research* 49(775), pp.697–741. March 1978.

² *The Mystical Life*, [Whm]; Whiteman lived 101 years, dying in February 2005, a few days after I happened to start reading his book.

³ [Cur], p.151.

⁴ [Mon], p.102.

⁵ First published in 1797 by Burton.

⁶ More about remote viewing in the last section of this chapter.

⁷ [Bro1], p.110.

⁸ [Jov], p.113.

⁹ [How], p.89.

¹⁰ [Whm], p.74.

¹¹ Carlos Castaneda: *Journey to Ixtlan* ([Cas]).

¹² [Cur], pp.117–118.

¹³ [Blac], p.124.

¹⁴ [Cot], p.63.

¹⁵ From Robert Crookall: *What Happens When You Die* ([Cro], p.56); the original source are tapes of interviews with Attila Von Szalay, given to Crookall by the writer Raymond Bayless.

[16] *Psicologi, ou Traité de l'apparition des espirits, á scavoir des âmes, séparées, fantasmes, ordiges, accidents merveilleux,* edition 160. See [deV], p.81.

[17] [Bla], p.87. Here is a very similar earlier experiment:

[Graham] Watkins was called in to investigate a haunted room in a Kentucky house. He brought with him a dog, cat, rat, and a rattlesnake to see how they would respond to the alleged haunting. The reactions were unusual and noteworthy. The dog, upon entering the room, snarled and backed out. It could not be induced to reenter the room. The cat reacted similarly and leaped from its owner's arms when being introduced into the room. It hissed and spat at the innocuous-looking unoccupied the chair in the centre of the room. The rat did not react at all, but the rattlesnake assumed a strike position focusing at the same chair as had the cat. None of these animals reacted in any manner in the control room. A tragedy (identified in the report) had occurred in the target room.

[18] [Tan], pp.125–126.

[19] Interview with Scott Rogo, in [Bla], p.73.

[20] [Rin2], p.109.

[21] [Kub], p.208.

[22] [Tal], p.242.

[23] [Grey], p.242.

[24] [Rin2], p.67.

[25] [Cro3], p.13, case 562.

[26] [Wei], pp.169–170.

[27] [Sab], pp.14–15.

[28] [Mul], p.114.

[29] [Mul], p.163.

[30] [Ste].

[31] [Ste], p.93.

[32] https://www.scientificexploration.org/docs/19/jse_19_1_keil. pdf. ([Kei]*).

[33] Source: https://www.pravda.ru/mysterious/46679-marsianin/.

[34] For example, [Mack].

[35] For example, [Can].

[36] [Wei], p.174.

[37] Also an article by Robert Aldemer in Gary Doore (Editor): *What Survives; Contemporary Explorations of Life After Death.*

[38] [Whi], pp.155–156.

[39] *Journal of the American Society of Psychical Research*, No. 4, October 1990; [Tar]*.

[40] [Tar].

[41] [Hor], pp.26-27.

[42] [Swe2].

[43] [Dus], p.133.

[44] [Cur], p.252.

[45] [Ste], p.117.

[46] [Wic], p.30.

[47] [Stev].

[48] [Smi1], pp.204–205.

[49] [Tart]*.

[50] [May], p.106.

[51] [May], p.107.

[52] [May], pp.10.

[53] [Swa]

[54] One recent example: https://boingboing.net/2020/04/07/what-are-these-mysterious-obje.html.

[55] [Can2], p.203.

Chapter 5

[1] [Fer].

[2] [Fer], p.60.

[3] Ibid., p.64.

[4] [Fer], p.142.

[5] Ibid., p.165.

[6] [Fon], p.157.

[7] [Sie].

[8] [Sie], p.17.

[9] Ibid., p.22.

[10] [Sie], p.30.

[11] [Str], p.155.

[12] Ibid., p.188.

[13] Ibid., p.189.

[14] Ibid., p.192.

[15] Ibid., p.198.

[16] [But], p.107.

[17] [Rog], p.20.

[18] [But], p.109.

[19] [Tar], p.361.

[20] Ibid., p.369.

[21] See, for example, [Mack].

[22] [Wal], p.63.

[23] [Val2], p76.

[24] Chronicler is Winsemius, year 1622.

[25] See Edgehill and Souter Fell: "A Critical Examination of Two English 'Phantom Army' Cases, by Peter A. McCue and Alan Gauld," *Journal of the Society of Psychical Research*, April 2005; and [Smy], p.43.

[26] [For], p.465.

[27] Ibid., p.466.

[28] April 5, 1897, edition of the Saginaw, Michigan "Evening News." (http://mysteriousuniverse.org/2011/09/anchors-aweigh-sky-ships-and-storm-wizards/)

[29] [Cano].

[30] [Wyl], p.78.

[31] [Ric], p.111. The original source is *Beyond Telepathy* by Andrija Puharić.

[32] [Del], p.181.

[33] [deV], p.171.

[34] [Spa2], p.64.

35 [Spa1], [Spa2], [Spa3], and [Spa4].
36 [Godf], p.60.
37 [Wat], p.202.
38 [Spa2], p.94.
39 As we will soon see, the phrase "objective observer" is an oxymoron from the scientific point of view.
40 [Iid], p.90.
41 Ibid, p.85.
42 [Ren], p.291.

Chapter 6

1 [Rhi], p.91.
2 [Tal], p.21.
3 [Myer], p.520.
4 [Spe], p.40.
5 [Tala], p.68.
6 [Smith], p.109.
7 [Owe], p.41.
8 [Holm], p.459.
9 [Jaf], pp.156–157.
10 [Dos], p.199.
11 [Targ2], p.87.
12 [Jac], p.53.
13 [Hon]*; [Rad2], p.165.
14 [Rad], p.161.
15 [McT], p.165.
16 Ibid.
17 The phrase is used by the entity Seth in Julia Roberts' books.
18 [Targ], p.58.
19 See, for example, [Ive], p.28.
20 [Snow].
21 [Rich], p.113; the original source is [Mye2], p.562.
22 [Fod], p.318.
23 [Pric], p.41.

[24] [Rich], p.167.

[25] [Fod], p.90.

[26] [Holm], p.98.

[27] [Fod], p.320.

[28] [SchS], p.70.

[29] Ibid., p.89.

[30] [SchS], p.74.

[31] Ibid., p.74.

[32] [Ryz]*, [Dup], [Ost].

[33] [Mac], p.108.

[34] [Stea], p.204.

[35] [Ran], p.61.

[36] [Mac], p.xi.

[37] [Ran], p.53.

[38] [Fan], p.132.

[39] [Fan], p.134.

[40] [Fan], p.135.

[41] [Ran], p.89.

[42] [Godf], p.247.

[43] There seems to be a strong connection between singularities in Earth's magnetic field and timeslips.

[44] http://sped2work.tripod.com/alaskabird0.html.

[45] [Heu], p.487.

[46] Ibid, pp.491–492.

[47] [Val2], p.185.

[48] [Laz], pp.331–333.

[49] [Mac], p.43.

[50] [Mac], p.52.

[51] [Ran], pp.65–69.

[52] [Ran], p.62.

[53] [Ran2], p.59.

[54] [Ran2], p.71.

[55] [Ran2], pp.76–77.

[56] [Ran2], p.90.

[57] [Don], pp.15–16.

[58] I know him well and I know that he will not mind this simplification.

[59] An article by Schmidt in Ramakrishna Rao (editor): *Basic Research in Parapsychology* ([Rao], Chapter 3). See also Schmidt's article "Observation of a Psychokinetic Effect Under Highly Controlled Conditions," in *Journal of Parapsychology*, Volume 57, December 1993.

[60] See Lynn Picknett's and Clive Prince's *The Forbidden Universe* ([Pic], p.319). The original article: https://www.researchgate.net/publication/6501301_Experimental_Realization_of_Wheeler%27s_Delayed-Choice_Gedanken_Experiment.

[61] All three of Monroe's very interesting books, [Mon1], [Mon2], and [Mon3], are required reading for anyone interested in OBE.

[62] [Snow].

[63] More support will be given in my next project; it is about time and space.

[64] [Ever], pp.30–31.

[65] Or we postpone it until we move in there.

Chapter 7

[1] [Spr], pp.81-82.

[2] Pine Point Trail, Whiteshell Provincial Park, Manitoba.

[3] [Rad2], p.197.

[4] Ibid.

[5] [Grof], pp.81–82.

[6] [Grof], pp.79–81.

[7] [Gur], p.81.

[8] See, for example, [Rhi2].

[9] See [Rad].

[10] See, for example, [Tag] or [Til].

[11] [Mos], p.126.

[12] On the other hand, the global internet and some more recent events seems to be reversing that trend.

[13] [Schw], p.60.

[14] [Fla5], p.416.

[15] [Mar], pp.154–155.

[16] [Olc], pp.76–77.

[17] [Wil], p.469.

[18] [DeM], p.26.

[19] [Bat]*.

[20] [Bro]*.

[21] [Das], p.380.

[22] [Hat], pp.211–215, Article by Bruce L. Cathie.

Chapter 8

[1] [Jaco], p.236.

[2] [Eva2], p.212.

[3] [Gros].

[4] [Fod], p.95.

[5] Ibid., pp.307–308; The original source is Lord Adare's *Experiences in Spiritualism with D. D. Home*, pp.155–157.

[6] [Mos].

[7] [Hat].

[8] [Str], p.199.

[9] [Brow], p.66.

[10] [Pet], p.78.

[11] [Cann].

[12] [Swam], pp.100–104.

[13] [Phi], p.137.

[14] [Hac], p.100; the original account was given by Ernesto Bozzano.

[15] [Phi], p.87.

[16] [Olc], pp.83–84.

[17] [Fod], p.14, and [Holm], pp.351, 353; the original source is *Rigid Tests of the Occult*, by "X" (Dr C. W. McCarthy of Sidney), *Light*, November 26, 1910.

[18] Via [Holm], pp.350–351.

[19] [Pric], p.89.

[20] [Gel], p.265.

[21] [Hac], p.80.

[22] [Fod], p.360.

[23] [Holm], p.361.

[24] [Bari], p.170.

[25] [Rie], p.4.

[26] [Stea], p.331.

[27] [Col], p.94.

[28] [Gre], p.73, or [Fod], p.100, or [Bard], p.161.

[29] [Owen], pp.317–318.

[30] [Gre], p.75.

[31] [Cro4], p.25.

[32] [Holm], pp.461–462.

[33] [Fla2], p.49, or [Batt], p.13.

[34] [Owen], p.333.

[35] [Fla2], p.60.

[36] [Spa1], pp.26–27.

[37] [Bra], p.25.

[38] [Bra], p.201.

[39] [Not], p.82.

[40] [Holm], p.370.

[41] [Pric], p.134.

[42] Ibid, p.131.

[43] [Pric], p.141.

[44] [Mar], pp.192–194.

[45] [Bai], p.273.

[46] Ibid.

[47] [Eas], p.82.

[48] Ibid.

[49] [Bra], p.207.
[50] [Fod], p.119.
[51] [Holm], p.435.
[52] [Croo], pp.219–220.
[53] [Fod], p.69.
[54] [Holm], p.398.
[55] Ibid., p.80.
[56] Ibid., pp.401–402.
[57] [Holm], p.402.
[58] Ibid., p.426.
[59] [Gis], p.79.
[60] [Gis], p.62.
[61] [Fod], p.116.
[62] [Das], p.384.
[63] [Fod], p.140.
[64] [Mar], p.10.
[65] [Fod], p.115.
[66] [Holm], p.439.
[67] [Cot], pp.70–72.
[68] [Bab], p.ii of the Appendix.
[69] [Bes].

Chapter 9

[1] Samuel Dill, *Roman Society from Nero to Marcus Aurelius*, MacMillan and Co., London, 1925, p.490.
[2] [Schw], p.175.

Epilogue

[1] [Ever], p.41.

Bibliography

The following items are referred to in the main text. The articles in the list are indicated by*.

[Bab]   Babbit, Edwin Dwight. *The Principles of Light and Color; Including Among Other Things the Harmonic Laws of the Universe, the Etherio-Atomic Philosophy of Force, Chromo Chemistry, Chromo Therapeutics, and the General Philosophy of the Fine Forces, Together with Numerous Discoveries and Practical Applications.* Babbitt & Co, 1878.

[Bai]   Baird, Alex. *Casebook for Survival.* Psychic Press Limited, London, 1947.

[Bard]   Bardens, Dennis. *Mysterious Worlds: A Personal Investigation of the Weird, the Uncanny, and the Unexplained.* Cowles Book Company, 1970.

[Bari]   Baring-Gould, Rev. S. *A Book of Folklore.* Singing Tree Press, 1970; facsimile reprint of the 1913 edition.

[Bar]   Barrett, Sir William. *Death-Bed Visions: the Psychical Experiences of the Dying.* Aquarian Press, Detroit, 1926; 1986 edition.

[Bat]*   Batcheldor, Kenneth James. "Report on a Case of Table Levitation and Associated Phenomena," *Journal of the Society of Psychic Research,* 1966.

[Batt]   Battersby, H.F. Prevost. *Man Outside Himself: The Facts of Etheric Projections.* Citadel Press, New Jersey, 1942.

[Bes]   Besant, Annie and Leadbeater, Charles W. *Occult Chemistry; Clairvoyant Observations on the Chemical Elements.* Theosophical Publishing House, 1919, third edition.

[Bla]   Black, David. *Ekstasy: Out-of-the-Body Experiences.* The Bobbs-Merrill Company Inc. Indianapolis/New York, 1975.

[Blac]   Blackmore, Susan J. *Beyond the Body; An Investigation of Out-of-the-Body Experiences.* A Paladin Book, 1982; First published by Heinemann, London, 1982.

[Blo]   Blood, Casey. *Science Sense & Soul: The Mystical-Physical Nature of Human Existence*. Renaissance Books, Los Angeles, 2002.

[Bra]   Bradley, H. Dennis. *The Wisdom of the Gods*. T. Werner Laurie Ltd., London, 1925.

[Bro]*   Brooks-Smith, C. and Hunt, D.W. "Some Experiments in Psychokinesis," *Journal of the Society of Psychical Research*, June 1970.

[Bro1]   Brown, Courtney. *Cosmic Voyage: Astonishing New Evidence of Extraterrestrials Visiting Earth*. Onyx Books, 1997.

[Brow]   Brown, Michael H. *A Report on the Powers of Psychokinesis, Mental Energy that Moves Matter*. Garber Communications, 1976.

[Bru]   Brunton, Dr Paul. *A Search in Secret Egypt*. BI Publications, Bombay, 1977 edition of the 1935 original.

[Bud]   Budge, Gavin. *Romanticism, Medicine and the Natural Supernatural: Transcendent Vision and Bodily Spectres, 1789–1852*. Palgrave Macmillan, London, 2012.

[But]   Buttlar, Johannes Von. *Journey to Infinity: Travels in Time*. Fontana/Collins, 1975; translation of the 1973 German edition.

[Cal]   Calmet, Dom Augustin. *Disertation upon Apparitions of Angels, Demons, and Spirits, Ghosts, and concerning the Vampires of Hungary, Bohemia, Moravia, and Silesia*. M. Cooper, London, 1746.

[Cano]   Cannon, Alexander. *The Powers That Be*. Kessinger Publishing, 1935.

[Can]   Cannon, Dolores. *The Convoluted Universe* (Books 1 to 5). Ozark Mountain Publishing.

[Can2]   Cannon, Dolores. *The Three Waves of Volunteers and the New Earth*. Ozark Mountain Publishing, 2011.

[Cann]   Cannon, Alexander. *The Powers That Be*. Kessinger Publishing, 1935.

[Car]   Cardan, Jerome (Girolamo Cardano; Hieronimus Cardanus). *The Book of My Life (De Vita Propria Liber)*, 1575. J. M. Dent and Sons, London, 1931; English edition.

[Carr]*   Carroll, Michael P. "Visions of the Virgin Mary: The
Effect of Family Structures on Marian Apparitions," *Journal
for the Scientific Study of Religion*, Volume 22, No. 3 (September
1983), pp.205–221.

[Cas]   Castaneda, Carlos. *Journey to Ixtlan: The Lessons of Don Juan*.
The Bronx, New York, 1982.

[Car]   Carrington, Hereward. *Psychic Oddities: Fantastic and Bizarre
Events in the Life of a Psychical Researcher*. Rider & Company,
1952.

[Chr]   Christion Jr, William A. *Apparitions in Late Medieval and
Renaissance Spain*. Princeton University Press, New Jersey, 1982.

[Col]   Colahan, Clark. *The Visions of Sor Maria de Agreda: Writing
Knowledge and Power*. 1994.

[Cot]   Cott, Jonathan. *In Search for Omm Sety: Reincarnation and
Eternal Love*. Doubleday, New York, 1987.

[Cro]   Crookall, Robert. *Intimations of Immortality: "Seeing" that
leads to "believing"*. James Clarke & Co., W.C., London, 1965.

[Cro2]   Crookall, Robert. *What Happens When You Die*. Collin
Smythe, Gerrards Cross, 1978.

[Cro3]   Crookall, Robert. *Case-Book of Astral Projection, 545–746*,
Citadel Press, Secaucus, New Jersey, 1972; Originally published
1874.

[Cro4]   Crookall, Robert. *More Astral Projections*. Aquarian Press,
London, 1965.

[Croo]   Crookes, William. *Crookes and the Spirit World: The Important Investigations by Sir William Crookes in the Field of Psychical
Research*. Taplinger Publishing Company, New York, 1972.

[Cur]   Currie, Ian. *You Cannot Die: The Incredible Findings of a Century of Research on Death*. Somerville House, 1998; First edition
1978.

[Das]   da Silva Mello, A. *Mysteries and Realities of This World and
the Next*. Weidenfeld & Nicolson, London, 1960; First published
in 1950.

[Del]   Deloria Jr, Vine. *The World We Used to Live in: Remembering the Powers of the Medicine Men*. Fulcrum Publishing, 2006.

[DeM]   De Morgan, Sophia Elizabeth (C.D.). *From Matter to Spirit: The Result of Ten Years' Experience in Spirit Manifestations*. Cambridge University Press, 1863.

[deV]   de Vesme, Caesar. *A History of Experimental Spiritualism: Volume 1, Primitive Man*. Rider & Co., London, 1931.

[Don]   Doncaster, Lucy and Holland, Andrew. *Greatest Mysteries of the Unexplained*. Arcturus Publishing Limited, London, 2020.

[Doo]   Doore, Gary (Editor). *What Survives: Contemporary Explorations of Life After Death*. Jeremy P. Tarcher, Inc., Los Angeles, 1990.

[Dos]   Dossey, Larry. *Healing Words: The Power of Prayer and the Practice of Medicine*. Harper, San Francisco, 1993.

[Dus]   van Dusen, Wilson. *The Presence of Other Worlds: The Psychological/Spiritual Findings of Emanuel Swedenborg*. Harper and Row, New York, 1974.

[Eas]   Eason, Cassandra. *The Psychic Powers of Children: Extraordinary experiences of ordinary families*. Rider, London, 1990.

[Eva]   Evans-Wentz, W.Y. *The Fairy Faith in Celtic Countries*. Oxford University Press, London, 1911.

[Eva2]   Evans-Wentz, W. Y. (Editor and Annotator). *Tibet's Great Yogi Milarepa: A Biography From the Tibetan Being the Jetsün-Kahbum or Bibliographica History of Jetsün-Milarepa, According to the Late Lama Kazi Dawa-Samdup's English Rendering*. Oxford University Press, New York, 1971, second edition; First edition 1951.

[Ever]   Evertz, Anke. *Nine Days of Eternity: An extraordinary near-death experience that teaches us about life and beyond*. Hay House, Carlsbad, California, 2022; Translation from German.

[Fan]   Fanthorpe, Lionel and Patricia. *Mysteries and Secrets of Time*. Dundurn Press, Toronto, 2007.

[Fer]   Fernandes, Joaquin and d'Armada, Fina. *Heavenly Lights, The Apparitions of Fatima and the UFO Phenomenon*. Anomalist Books, New York, 2007.

[Fla] Flammarion, Camille. *Death and its Mystery — Volume I: Before Death; Proofs of the Existence of the Soul.* The Century Co., New York and London, English translation 1923.

[Fla2] Flammarion, Camille. *Death and its Mystery — Volume 2: At the Moment of Death; Manifestations and Apparitions of the Dying; "Doubles," Phenomena of Occultism.* New York and London, English translation 1923.

[Fla3] Flammarion, Camille. *The Unknown.* Harper and Brothers, New York and London, 1900.

[Fla4] Flammarion, Camille. *Death and its Mystery — Volume 3: After Death; Manifestations and Apparitions of the Dead; "The Soul After Death."* New York and London, English translation 1922.

[Fla5] Flammarion, Camille. *Mysterious Psychic Forces.* Small Maynard and Co., Boston, 1907.

[Fod] Fodor, Nandor. *Encyclopedia of Psychic Science.* University Books, 1966; Third edition 1969; First published in 1934.

[Fon] de Fonseca, Père G. and Barthas, Chanoine C. *Our Lady of Light.* The Bruce Publishing Company, Milwaukee, 1947.

[For] Fort, Charles. *New Lands* (1923); within *The Collected Books of Charles Fort.* Henry Holt and Company, New York, 1974.

[Gel] Geley, Gustave. *Clairvoyance and Materialization.* T. Fisher Unwin Limited, London, 1927; English translation from French.

[Gis]* Gissurarson, Loftur R. and Haraldson, Erlandur. "The Icelandic Physical Medium Indridi Indridason," *Proceedings of the Society for Psychical Research,* Volume 57, Part 214, January 1989.

[Godf] Godfrey, Linda S. *Monsters Among Us.* A Tarcher Perigee Book, 2016.

[Gra] Granston, Sylvia and Williams, Carey. *Reincarnation: A New Horizon in Science, Religion and Society.* Julian Press, New York, 1984.

[Gre] Greenhouse, Herbert B. *The Astral Journey.* Avon Books, New York, 1974.

[Grey] Greyson, Bruce and Flynn, Charles. *The Near-Death Experience: Problems, Prospects, Perspectives.* C Thomas, Springfield, 1984.

[Grof]   Grof, Stanislav. *When the Impossible Happens, Adventures in Non-ordinary Realities*. Sounds True, 2006.

[Grof2]   Grof, Stanislav. *The Cosmic Game: Explorations of the Frontiers of Human Consciousness*. University of New York Press, Albany, New York, 1966.

[Gros]   Grosso, Michael. *The Man Who Could Fly: St. Joseph of Copertino and the Mystery of Levitation*. Rowman and Littlefield, 2016.

[Gug]   Guggenheim, Bill and Guggenheim, Judy. *Hello from Heaven: A new field of research — After-Death Communication confirms that life and love are eternal*. Bantam, New York, 1995.

[Gur]   Gurdjieff, George Ivanovich. *Meetings with Remarkable Men*. Picador/Pan Macmillan, 1963; 1978 edition.

[Hac]   Hack, Gwendolyn Kelley. *Modern Psychic Mysteries, Millesimo Castle, Italy*. Rider, 1929.

[Har]   Harlow, S. Ralph. *A Life After Death*. Doubleday, New York, 1961; 1968 edition.

[Hat]   Hatcher Childress, David. *Anti-Gravity and the World Grid*. Adventures Unlimited Press, Kempton, Illinois, 1987; Twelfth edition 2002.

[Heu]   Heuvelmans, Bernard. *On the Track of Unknown Animals*. Rupert Hart-Davis, 1958.

[Hil]   Hill, J. Arthur. *Man is a Spirit: A Collection of Spontaneous Cases of Dream, Vision and Ecstasy*. George H. Doran Company, New York, 1908.

[Holm]   Holms, A. Campbell. *The Facts of Psychic Science and Philosophy: Collated and Discussed*. University Books, New York, 1969; First published in 1925.

[Hon]*   Honorton, Charles and Ferari, Diane C. "Future-Telling: A Meta-Analysis of Forced-Choice Precognition Experiments," *Journal of Parapsychology*, Volume 53, December 1988, pp.281–309.

[Hor]   Horgan, John. *Rational Mysticism: Spirituality Meets Science In the Search For Enlightenment*. Houghton Mifflin Company, Boston, 2003.

[How] Howitt, William. *The History of the Supernatural (in all ages and nations and in all churches Christian and pagan demonstrating a universal faith)*, Volume 1. J. B. Lippincott & Co., Philadelphia, 1863.

[Ive] Iverson, Jeffrey. *More Lives Than One: The Bloxham Tapes*. Pan Books, London and Sidney, 1977.

[Jac] Jacobsen, Nils O. *Life Without Death?* Dell Publishing Co., 1971.

[Jaco] Jacolliot, Louis. *Occult Science in India and Among the Ancients*. Cosimo Classics, 1884; 2005 edition.

[Jaf] Jaffé, Aniela. *Apparitions: An Archetypal Approach to Death Dreams and Ghosts*. Spring Publications Inc., Irving, Texas, 1957; 1963 translation from German.

[Jov] Jovanovic, Pierre. *An Inquiry into the Existence of Guardian Angels: A Journalist's Investigating Report*. M. Evans & Company, New York, 1993; This edition 1995.

[Jun] Jung, Carl. *Memories, Dreams, Reflections*. (Recorded and edited by Aniela Jaffe). Pantheon Books, New York, 1963.

[Keel] Keel, J.A., *The Mothman Prophecies*. Saturday Review Press, 1975.

[Kei]* Keil, H.H. Jürgen and Tucker, Jim B. "Children Who Claim to Remember Previous Lives: Cases with Written Records Made before the Previous Personality Was Identified," *Journal of Scientific Exploration*, Volume 19, No. 1, pp.91–101, 2005.

[Kes] Kessler, David. *Vision, Trips, and Crowded Rooms: Who and What You See Before You Die*. Hay House, Carlsbad, California, 2010.

[Kni] Knight, David C. (Editor). *The ESP Reader*. Grosset & Dunlop, New York, 1969.

[Kub] Kübler-Ross, Elisabeth. *On Children and Death*. Macmillan, London, 1983.

[Laz] Lazarus, Richard. *Beyond the Impossible: A Twentieth Century Almanac of the Unexplained*. Warner Books, London, 1994.

[Lun]   Lundahl, Craig Ph.D. and Widdison, Harold Ph.D. *The Eternal Journey: How Near–Death Experiences Illuminate Our Earthly Lives*. Warner, New York, 1997.

[Mack]   Mack, John E. *Abduction, Human Encounters with Aliens*. Simon & Schuster, 1995.

[Mac]   MacKenzie, Andrew. *Adventures in Time: Encounters with the Past*. Athlone P., 1997.

[Mae]   Maeterlinck, Maurice. *The Unknown Guest*. University Books, Secaucus, New Jersey, 1974.

[Mar]   Marryat, Florence. *There Is No Death*. National Book Company, New York, 1917; 1938 edition.

[May]   Mayer, Elizabeth Lloyd Ph.D. *Extraordinary Knowing: Science, Skepticism, and the Inexplicable Powers of the Human Mind*. Bantam, 2007.

[McC]   McCarthy-Jones, Simon. *Hearing Voices: The History and Meaning of Auditory Verbal Hallucinations*. Cambridge University Press, California, 2012.

[McT]   McTaggart, Lynne. *The Intention Experiment: Using Your Thoughts to Change Your Life and the World*. Free Press/Simon and Schuster, 2007.

[Mon1]   Monroe, Robert A. *Journeys out of the Body*. Anchor Books, 1971.

[Mon2]   Monroe, Robert A. *Far Journeys*. Doubleday, 1985.

[Mon3]   Monroe, Robert A. *Ultimate Journey*. Harmony, 1994.

[Mon]   Montgomery, Ruth. *Aliens Among Us*. Fawcett Crest, New York, 1985.

[Moo]   Moody, Raymond with Paul Perry. *Glimpses of Eternity*. Guideposts, New York, 2010.

[More]   More, Henry. *The Immortality of the Soul*. Springer, Martinus Nijhoff Publishers, Boston, 1642; 1987 edition edited by A. Jacob.

[Mor]   Morse, Melvin M.D. and Paul Perry. *Closer to the Light: Learning from the near-death experiences of children*. Raymond Moody & Paul Perry, 1990.

[Mor2]   Morse, Melvin with Paul Perry. *Parting Visions: Uses and Meaning of Pre-Death Psychic and Spiritual Experiences*. Villard Books, New York, 1994.

[Mor]* Morse, M., Castillo, P., Venecia, D. *et al.* "Childhood Near-Death Experiences," *American Journal of Diseases of Children* 140 (1986), pp.1110–1113.

[Mos]   Moss, Thelma. *The Probability of the Impossible: Scientific Discoveries and Explorations in the Psychic World*. Routledge & Kegan Paul, London, 1976.

[Mul]   Muller, Karl E. *Reincarnation – Based on Facts*. Psychic Press Ltd., London, 1970.

[Mye]   Myers, Arthur. *The Ghostly Register*. McGrow Hill, 1986.

[Myer]   Myers, Frederic W.H. *Human Personality and its Survival of Bodily Death*, Volume I. Longmans, Green, and Co., London, 1907.

[Mye2]   Myers, Frederic W.H. *Human Personality and its Survival of Bodily Death*, Volume II. Longmans, Green, and Co., London, 1903; 1954 edition.

[Not]   Notzing, Albert Von Schrenck. *Phenomena of Materialization: A Contribution to the Investigation of Mediumistic Teleplastics*. Kegan, Trench, Trubner & Co., London, 1913.

[Oat]   Oaten, Ernest W. *That Reminds Me: A Medley of Personal Psychic Experiences*. Worlds Publishing Company, 1938.

[Olc]   Olcott, Henry Steel. *Isis in America*. Jeremy P. Tarcher, Penguin, 2014; the original title was *Old Diary Leaves* and it was published 1895.

[Ose]   Ossendowski, Ferdinand. *Beasts, Men and Gods*. Dutton, New York, 1923.

[Osi]   Osis, Karlis. *Deathbed Observations by Physicians and Nurses*. Parapsychology Foundation, Inc., New York, 1961.

[Ost]   Ostrander, Sheila and Schroeder, Lynn. *Psychic Discoveries Behind the Iron Curtain*. Prentice Hall, 1971, fifth printing; First edition 1970.

[Owen]   Owen, Robert Dale. *Footfalls on the Boundary of Another World*. Cambridge University Press, 1860.

[Owe]   Owen, A.R.G. *Psychic Mysteries of Canada*, Fitzhenry & Whiteside, Toronto. 1975.

[Pete]   Peters, William J. and Kinsella, Michael. *At Heavens Door: What Shared Journeys to the Afterlife Teach About Dying Well and Living Better*. Simon and Schuster, New York, 2022.

[Pet]   Peterson, Joseph H. (Editor). *John Dee's Five Books of Mystery: Original Sourcebook of Enochian Magic*. Weiser Books, 2003.

[Phi]   Phillips, P. and S. *Is Death the End: A Unique Psychic Experience*. Corgi Books, 1971.

[Pic]   Picknett, Lynn and Prince, Clive. *The Forbidden Universe*. Constable, London, 2011.

[Pop]   Pope, Nick. *The Uninvited: An Expose of the Alien Abduction Phenomenon*. Simon & Schuster, 1997.

[Pric]   Price, Harry. *Fifty Years of Psychical Research*. Longmans, Green and Co., London, 1939; 1975 edition.

[Rad]   Radin, Dean. *Supernormal: Science, Yoga, and the Evidence for Extraordinary Psychic Abilities*. Deepak Chopra Books, 2013.

[Rad2]   Radin, Dean. *Entangled Minds: Extrasensory Experiences in a Quantum Reality*. Simon & Schuster, 2006.

[Ran]   Randles, Jenny. *Time Travel: Fact, Fiction & Possibility*. Blandford Press, London, 1994.

[Ran2]   Randles, Jenny. *Time Storms: Amazing Evidence for Time Warps, Space Rifts and Time Travel*. Judy Piatkus Publishers, London, 2001.

[Rao]   Rao, Ramakrishna (Editor). *Basic Research in Parapsychology*. McFarland & Co., 2001.

[Ree]   Reed, Graham. *The Psychology of Anomalous Experience: A Cognitive Approach*. Hutchison University Library, London, 1972.

[Rei]   Reilly, Carmel. *True Tales of Angel Encounters*. Llewellyn Publication, Woodbury, Minnesota, 2009.

[Ren]   Renard, Gary R. *The Disappearance of the Universe*. Fearless Books, Berkeley, California, 2003.

[Rhi] Rhine, Louisa. *The Invisible Picture: A Study of Psychic Experiences*. McFarland, Jefferson, NC, 1981.

[Rhi2] Rhine, Louisa E. *Mind over Matter — Psychokinesis: the astonishing story of the scientific experiments that demonstrate the power of the will over matter*. Collier McMillan Publishers, London, 1970.

[Rich] Richet, Charles. *Our Sixth Sense*. Rider & Co., London, 1925; Translation from French.

[Ric] Rickard, Bob and Michell, John. *The Rough Guide to Unexplained Phenomena*. Rough Guides, London, New York, 2007.

[Rie] Riedweg, Christoph. *Pythagoras*. Cornell University Press, 2002; 2005 translation from German to English.

[Rin] Ring, Kenneth. *Heading Toward Omega: In Search of Meaning of the Near-Death Experience*. W. Morrow, 1985.

[Rin2] Ring, Kenneth. *Lessons from the Light*. Da Capo Press, University of California, 1998.

[Rog] Rogo, D. Scott. *Leaving the Body: A Complete Guide to Astral Projections —A step-by-step presentation of eight different systems of out-of-body travel*. Prentice Hall, New Jersey, 1983.

[Ros]* Rosenhan, D. L. On "Being Sane in Insane Places," *Science*, Volume 179, January 1973.

[Ryz]* Ryzl, Milan and Pratt, J.G. "A Further Confirmation of a Stabilized ESP Performance in a Selected Subject," *The Journal of Parapsychology*, Volume 27, June 1963, pp.73–83.

[Sab] Sabom, Michael B. *Recollection of Death: A Medical Investigation*. Harper & Row, 1982.

[Sac] Sacks, Oliver. *Hallucinations*. Alfred A. Knopf, New York-Toronto, 2012.

[Schw] Schwartz, Gary E. *The Sacred Promise: How Science is Discovering Spirit's Collaboration With Us in Our Daily Lives*. Atria Books, New York, 2011.

[SchS] Schwartz, Stephan A. *The Secret Vaults of Time; Psychic Archeology and the Quest for Man's Beginnings*. Grosset & Dunlap, New York, 1978.

[Sco] Scott, Cyril (The Pupil). *The Initiate; Some Impressions of a Great Soul*. Samuel Weiser Inc., 1981; First published in 1920.

[She] Sheldrake, Rupert. *The Sense of Being Stared at: and Other Aspects of the Extended Mind*. Crown Publishers, New York, 2003.

[Sie] Siegel, Ronald K. *Fire in the Brain, Clinical tales of Hallucination*. Plume, New York, 1992.

[Sla]* Slawinski, Janusz. "Electromagnetic radiation and the afterlife," *Journal of Near-Death Studies*, December 1987, Volume 6, pp.79–94.

[Smith] Smith, Alson J. *Immortality: The Scientific Evidence*. Prentice Hall, New Jersey, 1954.

[Smi1] Smith, Frederic M. *The Self Possessed: Deity and Spirit Possession in South Asian Literature and Civilization*. Columbia University Press, 2006.

[Smit] Smith, Robert C. *In the Presence of Angels*. A.R.E. Press, Virginia Beach, 1993.

[Smy] Smyth, Frank and Stemman, Roy. *Mysteries of Afterlife*. Ferguson Publishing Company, 1991.

[Snow] Snow, Chet. *Mass Dreams of the Future* (Featuring, *Hypnotic future-life Progressions* by Helen Wambach). McGrow-Hill, New York, 1989.

[Spa1] Spalding, Baird T. *Life and Teaching of the Masters of the Far East*, Volume 1. De Vorss & Co., 1964; 1924.

[Spa2] Spalding, Baird T. *Life and Teaching of the Masters of the Far East*, Volume. 2. De Vorss & Co., 1964; 1924.

[Spa3] Spalding, Baird T. *Life and Teaching of the Masters of the Far East*, Volume 3. De Vorss & Co., 1962; 1935.

[Spa4] Spalding, Baird T. *Life and Teaching of the Masters of the Far East*, Volume 4. De Vorss & Co., 1948.

[Spe] Spees, Jenifer (Editor). *Mystic Experiences*. Barnes and Noble, 2001.

[Spr] Spraggett, Allen. *New Worlds of the Unexplained*. The New American Library of Canada Limited, 1976.

[Stea] Stead, W.T. *Borderland: A Casebook of True Supernatural*

Stories. University Books, New York, 1970.

[Ste]   Steiger, Brad. *Real Ghosts, Restless Spirits, and Haunted Places*. Visible Ink Press, Detroit, 2003.

[Stev]   Stevens, E.W. *Watseka Wonder*. Religio-Philosophical Publishing House, Chicago, 1878.

[Ste]   Stevenson, Ian. *European Cases of the Reincarnation Type*. McFarland, Jefferson, NC, 2003.

[Str]   Strassman, Rick. *DMT: The Spirit Molecule*. Park Street Press, Rochester, Vermont, 2001.

[Stra]   Strassman, Rick; Wojtowicz, Slawek; Luna, Luis Eduardo and Frecska, Ede. *Inner Paths to Outer Space: Journeys to Alien Worlds through Psychedelics and Other Spiritual Technologies*. Park Street Press, Rochester, Vermont, 2008.

[Str]   Strieber, Whitley and Kripal, Jeffrey J. *The Supernatural: A New Vision of the Unexplained*. Jeremy Tarcher / Penguin, New York, 2016.

[Stre]   Streeter, B.H. and Appasamy, A.J. *The Sadhu: A Study in Mysticism and Practical Religion*. MacMillan and Co., London, 1927.

[Swe2]   Swedenborg, Emanuel. *Arcana Coelestia (the Heavenly Arcana)*, Volume XI, 9112–9973, 1749–1756.

[Swam]   Rama, Swami. *Living With the Himalayan Masters*. Himalayan International Institute of Yoga Science & Philosophy of the USA, Honesdale, PA, 1979.

[Swa]   Swann, Ingo. *Penetration: The Question of Extraterrestrial and Human Telepathy*. Ingo Swann Books (samizdat – seems typed on a typewriter), 1998.

[Tag]   McTaggart, Lynne. *The Intention Experiment: Using Your Thoughts to Change Your Life and the World*. Atria Books, 2007.

[Tala]   Talamonti, Leo. *Forbidden Universe: Mysteries of the Psychic World*. Stein and Day, New York, 1975.

[Tan]   Tanous, Alex with Harvey Ardman. *Beyond Coincidence: One Man's Experiences With Psychic Phenomena*. Doubleday & Company, Inc., Garden City, New York, 1976.

[Targ]   Targ, Russell and Harary, Keith. *The Mind Race: Under-standing and Using Psychic Abilities*. Random House, 1984.

[Targ2]   Targ, Russell. *Limitless Mind: A guide to remote viewing and transformation of consciousness*. New World Library, Novato, 2004.

[Tart]*   Tart, Charles. "Psychophysiological Study of Out-of-the-Body Experiences, in a Selected Subject," *Journal of the American Society for Psychical Research*, 1968.

[Tart]   Tart, Charles (Editor). *Altered States of Consciousness*. John Wiley & Sons, 1969; 1990 updated edition.

[Tar]   Tarazi, Linda. *Under the Inquisition: An Experience Relived*. Hampton Roads, 1997.

[Tar]*   Tarazi, Linda. "An Unusual Case of Hypnotic Regression with Some Unexplained Content," *Journal of the American Society of Psychical Research*, No. 4, October 1990.

[Til]   Tiller, William A.; Dibble, Walter E.; Kohane, Michael J. *Conscious Act of Creation: An Emergence of a New Physics*, Pavior Pub., 2001.

[Tro]   Trobridge, George. *Swedenborg, Life and Teaching*. Swedenborg Foundation, New York, 1962; Fifth reprint of the fourth edition, 1935.

[Tuc]   Tucker, Jim B. *Life Before Life*. Piatkus, New York, 2006.

[Val]   Vallee, Jacques. *Revelations: Alien Contact and Human Deception*. Ballantine Books, New York, 1991.

[Val2]   Vallee, Jacques and Aubeck, Chris. *Wonders in the Sky: Unexplained Aerial Phenomena From Antiquity to Modern Times*. Jeremy P. Tarcher/ Penguin, New York, 2009.

[Wal]   Walsh, Roger and Grob, Charles S. (Editors). *Higher Wisdom: Eminent Elders Explore the Continuing Impact of Psychedelics*. State University of New York Press, Albany, 2005.

[Wan]   Wang, Hao. *A Logical Journey: From Gödel to Philosophy*. The MIT Press, Cambridge, Massachusetts, 1996.

[Wat]   Watson, Lyall. *Gifts of Unknown Things*. Simon and Schuster, New York, 1976.

[Wei]   Weiss, Brian. *Messages from the Masters: Tapping into the Power of Love*. Piatkus Books, New York, 2007.

[Whm]   Whiteman, J.H.M. *The Mystical Life*. Faber and Faber, London, 1961.

[Whi]   Whitton, Joel L. and Fisher, Joe. *Life between Lives*. Garden City, New York, 1986.

[Wic]   Wickland, Carl A. M.D. *30 Years Among the Dead*. Coles, around 1924; 1980 edition.

[Wils] Wilson, Ian. *The After Death Experience*. Sidgwick and Jackson Ltd., 1987.

[Wil]   Wilson, Colin. *The Occult: A History*. Random House, New York, 1971.

[Wyl]   Wyllie, Timothy. Confessions of a Rebel Angel: The Wisdom of the Watchers and the Destiny of Planet Earth. Bear and Company, Rochester Vermont — Toronto, Canada, 2012.

[Yag]   Yagan, Murat. I Came From Behind Kaf Mountain: The Spiritual Biography. Threshold Books, Putney, Vermont, 1984.

6TH
BOOKS

ALL THINGS PARANORMAL

Investigations, explanations and deliberations on the paranormal, supernatural, explainable or unexplainable. 6th Books seeks to give answers while nourishing the soul: whether making use of the scientific model or anecdotal and fun, but always beautifully written.
Titles cover everything within parapsychology: how to, lifestyles, alternative medicine, beliefs, myths and theories.
If you have enjoyed this book, why not tell other readers by posting a review on your preferred book site?

Recent bestsellers from 6th Books are:

The Scars of Eden
Paul Wallis
How do we distinguish between our ancestors' ideas of God
and close encounters of an extraterrestrial kind?
Paperback: 978-1-78904-852-0 ebook: 978-1-78904-853-7

The Afterlife Unveiled
What the dead are telling us about their world!
Stafford Betty
What happens after we die? Spirits speaking through mediums
know, and they want us to know. This book unveils their
world...
Paperback: 978-1-84694-496-3 ebook: 978-1-84694-926-5

Harvest: The True Story of Alien Abduction
G. L. Davies
G. L. Davies's most-terrifying investigation yet reveals one
woman's terrifying ordeal of alien visitation, nightmarish
visions and a prophecy of destruction on a scale never before
seen in Pembrokeshire's peaceful history.
Paperback: 978-1-78904-385-3 ebook: 978-1-78904-386-0

Wisdom from the Spirit World
Carole J. Obley
What can those in spirit teach us about the enduring bond of
love, the immense power of forgiveness, discovering our life's
purpose and finding peace in a frantic world?
Paperback: 978-1-78904-302-0 ebook: 978-1-78904-303-7

Spirit Release
Sue Allen
A guide to psychic attack, curses, witchcraft, spirit attachment, possession, soul retrieval, haunting, deliverance, exorcism and more, as taught at the College of Psychic Studies.
Paperback: 978-1-84694-033-0 ebook: 978-1-84694-651-6

Advanced Psychic Development
Becky Walsh
Learn how to practise as a professional, contemporary spiritual medium.
Paperback: 978-1-84694-062-0 ebook: 978-1-78099-941-8

Where After
Mariel Forde Clarke
A journey that will compel readers to view life after death in a completely different way.
Paperback: 978-1-78904-617-5 ebook: 978-1-78904-618-2

Poltergeist! A New Investigation into Destructive Haunting
John Fraser
Is the Poltergeist "syndrome" the only type of paranormal phenomena that can really be proven?
Paperback: 978-1-78904-397-6 ebook: 978-1-78904-398-3

A Little Bigfoot: On the Hunt in Sumatra
Pat Spain
Pat Spain lost a layer of skin, pulled leeches off his nether regions, and was violated by an Orangutan for this book.
Paperback: 978-1-78904-605-2 ebook: 978-1-78904-606-9

Astral Projection Made Easy
and overcoming the fear of death
Stephanie June Sorrell
From the popular Made Easy series, Astral Projection Made Easy helps to eliminate the fear of death through discussion of life beyond the physical body.
Paperback: 978-1-84694-611-0 ebook: 978-1-78099-225-9

Haunted: Horror of Haverfordwest
G. L. Davies
Blissful beginnings for a young couple turn into a nightmare after purchasing their dream home in Wales in 1989.
Paperback: 978-1-78535-843-2 ebook: 978-1-78535-844-9

Readers of ebooks can buy or view any of these bestsellers by clicking on the live link in the title. Most titles are published in paperback and as an ebook. Paperbacks are available in traditional bookshops. Both print and ebook formats are available online.

Find more titles and sign up to our readers' newsletter at
www.6th-books.com

Join the 6th books Facebook group at
6th Books The world of the Paranormal